AF559703

Principles of Cytogenetics

Principles of Cytogenetics

K.P.S. Malik
Professor
Department of Plant Breeding & Genetics
College of Agriculture
Rajmata Vijayaraje Scindia Krishi Vishwavidyalaya
Gwalior, Madhya Pradesh

NEW INDIA PUBLISHING AGENCY
New Delhi – 110 034

NEW INDIA PUBLISHING AGENCY
101, Vikas Surya Plaza, CU Block, LSC Market
Pitam Pura, New Delhi 110 034, India
Phone: + 91 (11)27 34 17 17 Fax: + 91(11) 27 34 16 16
Email: info@nipabooks.com
Web: www.nipabooks.com

Feedback at feedbacks@nipabooks.com

ISBN No. 978-93-87973-99-2

Composed and Designed by NIPA

Preface

The book is basically intended to accompany course in cytogenetics for students of Genetics and Plant Breeding in Agricultural universities. Present book with the help of diagrams and explanations it has been attempted that even a beginner could grasp the core elements of the subject.

The book has been strictly organized on the basis of course curriculum of post graduate and PhD programs especially for courses "Principles of Cytogenetics" (GP502) and "Molecular and Chromosomal Manipulation for Crop Breeding" (GP604) being taught in Agricultural Universities. This book also covers majority of the course outlines of course of Genetics for B Sc and BSc(Ag.) students. All the topics covered in the book have been ordered in a crisp and comprehensible manner avoiding complexities of a traditional textbook since it is a simply a guide book to supplement but not supplant the main texts. Moreover, this book is also open one for genuine suggestions, comments and corrections from any source/s which would be duly acknowledged and incorporated in the present text. The main objective of the book is just to help the students to prepare themselves for examination and competition in a quick way in short period.

I would like to thank the many people without whose help or influence this book would not have been possible. Thanks to Prof. P.K.Gupta, my teacher and mentor who taught me the alphabet of this subject which helped me throughout my service period from assistant professor to Faculty Dean of RVSKVV, Gwalior(MP).Thanks to Prof. A.K Singh, Director Instruction, RVSKVV, Gwalior, Prof. A.K Sharma Deptt. of Plant Breeding and Genetics and Prof.(Smt) Reeti Singh who suggested me to put up my scrawled lecture notes in order and be published. I also extend my thanks to Mr. Sarman Karosiya and Rajesh Pal (computer operator) who helped immensely to prepare the manuscript on computer. Finally thanks to my wife Usha Malik and childern Ajeeta Singh, Ira Malik and Abhijeet Singh who always motivated me to focus on academic persuits. Last but not least I would extend my thanks to publisher, New India Publishing Agency, New Delhi for expressing his keen interest to publish this book.

Author

Contents

1

Morphology of Eukaryotic Chromosome

Most of the DNA of an living organism is packaged and organized in a chromosome as well as in mitochondria and cell organelles (Feulgen and Rossenbeck 1924). DNA is not found on its own but rather is structured in long strands wrapped around protein complex called **nucleosome** consisted of a protein called **histone**. During most of the duration of cell cycle a chromosome consists of long double helix DNA molecule (with associated proteins)During S phase the chromosome gets replicated resulting in X –shaped structure (chromatids) and still attached to a single centromere. The basic role of the chromosome is to provide a frame work which allows each linear segment of the genome to replicate and segregate efficiently. Failure in either of processes cause chromosome imbalance in daughter cells. Three specific cis-acting sites are required for stable chromosome maintenance, i.e., an origin of replication, a centromere and telomeres.

Classical studies of the morphology of individual chromosomes of the plant species begun by S. Nawaschin 1910-16 and continued by Taylor (1925), Sharp (1929) and Mc Clintock (1929-30). And eventually it was possible to set up working models which account for the internal structure and chemistry of the chromosome and the genetic material as well as genetical facts. The use of aceto-carmine with iron (Belling, 1921) and application of heat greatly improved the staining differentiation of chromosomes. Smears have practically replaced the paraffin techniques of studying chromosome. The gross morphology of chromosomes is generally worked out either at prophase stages of meiosis or metaphase in somatic tissues. The chromosomes may differ in thickness, total length, position of centromere or spindle fiber attachment region, chromomere pattern and number and position of secondary constriction. The total length of chromosomes at metaphase also differs in different tissues of a species and also in certain tissues chromosome present may be very different in appearance. Figure 1 shows various morphological features of a eukaryotic chromosome

1. Centromere and position of the centromere

The centromere or primary constriction is known by various other names as spindle fibre attachment region (SFA), kinetochore and insertion region. Usually the centromere in a chromosome is not terminal and hence it divides the chromosome in two arms, i.e., a and b. The ratio of long/short (a/b) arms lengths is relatively constant for a particular chromosome. In pachytene chromosomes in maize, the centromere is a short and very lightly stained region (usually invisible) which differs in length for different chromosomes but constant for a particular chromosome. The polycentric or diffuse condition of centromere may also exist in certain species like coccids and plant species *Luzula purpurea* (woodrush). Centromere (kinetochore) is the most conspicuous (easily noticed) feature of most chromosomes, appearing from mitotic prophase to anaphase as region which does not coil. It is weakly stained and can be distinguished from darkly stained chromosome arms. The distinctness of centromere varies a great deal between various organisms and it can be enhanced further by pretreatment with mitotic inhibitors like paradichlorobengene before staining. Although position of the centromere is constant for a given chromosome but it can vary between different chromosomes and so they are categorized on the basis of the centromere location as acrocentric, metacentric and telocentric. Usually the term kinetochore and centromere are used synonymously but presently the term kinetochore is secured for internal structure within centromere by which chromosomes are moved towards poles during cell division. lima-de-Faria (1949)showed that centromere contained 2 to 5 pairs of darker staining chromomeres joined by uncoiled chromonematal fibrils.

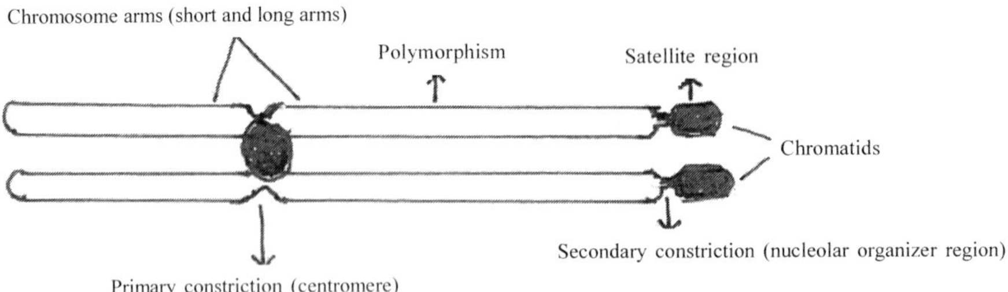

Fig. 1: Morphological features of a eukaryotic metaphase chromosome. The centromere or primary constriction shows position of chromosome, stains densely and joins all four chromatids. Secondary constrictions are lightly colored and represent nucleolar organizer region. Telomeric chromosome segments located distal to nucleolar organizer region are termed satellite region. Nucleolar organizers commonly contribute to nucleolus. Chromosome polymorphisms /heteromorphism are heritable morphological features of a chromosome which vary within a population having no phenotype. It includes areas of variable heterochromatin, inversions and fragile sites.

However, electron microscopy of sectional material shows that practically all metaphase chromosomes can be interpreted as having one or less circular kinetochore per chromatid. In species of *Tradiscantia, Rhaeo and Allium* the kinetochore when stained by Giemsa technique during mid prophase to mid metaphase. They are about 0.5 μm in diameter and sometimes appear to be attached to remanants of spindle fiber. By metaphase the two kinetochoes, one per chromatid; are clearly separated and lie laterally on opposite sides of centromere.The kinetochore is believed to be rich in repeatitive DNA, which would permit its division; as suggested by observations on chromosomal structural changes.It would seem that no more than two of Lima-de Faria "centromeric chromomeres"can be kinetochores Lima –de Faria(1949) showed that the centromere contained two to five pairs of darker staining chromeres joined by uncoiled chromonemetal fibrils (Fig.2).

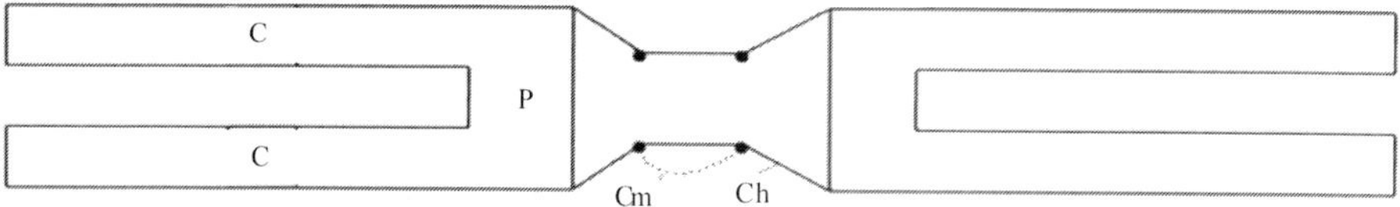

Fig. 2: Redrawn centromere model of Lima-de Faria (1949). Pairs of centromeric chromomeres (cm) are connected to each other and to the proximal regions(P) joining the sister chromatids(C) by the chromatonemal fibrils.

In some organisms the centromere is unlocalized so that many points along the chromosome can function as kinetochore and it has been shown that diffuse or polycentric chromosomes of, for example, *Luzula and Cyperus* do not possess kinetochores of the kind described above.

2. Telomeres

Telomeres are specialized nucleoprotein structures at the ends of chromosomes to protect its integrity.They are reported to get shorter each time a cell copies itself but keeping the important DNA intact.If chromosomes are fractured by X- ray, for example, the resulting segments may fuse again; however,they will not fuse with the ends of chromosome arms. This has led to the conclusion that a chromosome is terminated by telomere which confirms polarity upon it.The telomere has been shown to be a compound structure consisting of several differetiated segments; in rye, for example, it is composed of atleast two pairs of chromomeres and intercalary fibrils. Breakage within this region should still give stable chromosome ends as has been demonstrated in rye and maize. The compound structure of telomere is very like that of the centromere and they are further shown to share a number of properties relating to their cycle of division and behaviour during meiosis.

3. Secondary constriction and satellites

In many chromosomal complements atleast one pair of chromosomes is seen to have an unspiralized non-staining region in addition to the centromere.This secondary constriction frequently occurs near the end of a chromosome so that the segment beyond the constriction is small which is termed as satellite or trabant joined by the rest of chromosome by a stalk known as **satellite stalk** (Figure 1).

The only known function of secondary constriction is that of nuleolar organization and it is believed that the nuleoli (singular-nuleolus) involved in protein synthesis (specially contains genes for r RNA synthesis) are controlled by specific loci associated with secondary constriction. In maize nuleolar organizing element is at the base of satellite stalk and it has been shown that following its fragmentation by x-irradiation, both subunits remain functional (satellite and nucleolus) thus indicating it to be a compound structure as centromere and telomere.

The number of satellited chromosomes in the complement varies in different organisms and is not always parallelled by the number of nucleoli visible at prophase.Similarly chromosomes associated with nuleolar organization may not possess satellites, as in *Nothocordum inutile*, for example, in four chromosomes concerned with this activity seem to have compound structure which plays the role of both nucleolar oraganizer and centromere.

Furthermore, chromosomes are characterized as satellited or non-satellied ones based on presence or absence of the satellite. However, secondary constriction can vary greatly during course of mitosis as during prometaphase the satellites are joined to the chromosome arm by long slender satellite stalk which might readily be fractured during squash preparation while at metaphase particularly following pretreatment with drugs such as oxyquinoline, the stalk might be so short as to make it difficult to determine the presence of satellite, further more homologous chromosomes may differ in size of their satellites. The nucleolar organizer of one species may be dominant to that of another species so that, in hybrids the chromosomes of the weaker set may not show the satellite present in the parent.

4. Euchromatin and heterchromatin

The standard sequence of chromosomal condensation (at its maximum during metaphase –anaphase) and elongation (greatest) during interphase for the chromonemata during nuclear cycle is known as eucycle and the chromatin which follows this sequence is termed **euchromatin**.However some chromosome segments, and even whole chromosomes, have a different cycle so that they are more condensed and deeper stained than remainder of the complement at prophase (positively heteropycnotic) and negatively heteropycnotic later in the nuclear division when they appear under-condensed and lightly stained such allocyclic chromosomes or segments are described as **heterochromatic** and their occurrence and position provide useful markers in charting the kayotypes of a population or species.Except heteropycnosis, the heterochromatin differs from euchromatin in its behavior and, perhaps, structure, since it is inactive in transcription and contains late replicating, in some cases definitely repetitive DNA. Heterochromatin is classified in two classes, i.e., Constitutive and facultative heterochromatin. Regions of constitutive heterochromatin regularly show the feature described above and have few or no structural genes while facultative heterochromatin contains structural genes; cytologically condensed and genetically inactive and occuring in particular tissues being apparently dependent upon certain physiological and developmental conditions.

1. Types of chromosomes

(A) B-chromosomes

Each cell of eukaryotic normal individual carries a well defined set of chromosomes (karyotype) which is exploited as a diagnostic tool of the species. However, some species carry extra chromosomes which are composed of heterochromain. These chromosomes have no phenotyptic effect and vary between populations and even between cells of an individual . They are referred to as B chromosomes (plants) or accessory chromosomes, satellite chromosomes and supernumerary chromosomes. They are reported to have been derived from A-complement following reciprocal translocation as in case of *Clarkia elegans* or they may originate denovo in populations with higher levels of translocation heterozyosity. But in majority of the cases their origin is not definite.

(B) Meiosis in of B-chromosomes

During meiosis homologous B-chromosomes can pair forming multivalent associations and chiasma frequency between them is highly subject to environmental factors. B bivalents normally orientate and segregate normally at meiosis. However unpaired ones show variable behavior at first anaphase like univalents. In many grasses they have been reported to divide consistently but during anaphase second daughter B-chromosomes lag on spindle and thus are excluded from tetrad nuclei. But in other cases such as *Anthoxantham aristatum,* they divide only at anaphase II and segregate normally to the tetrad. During mitosis they may divide normally or show various kinds of instability such as non disjunction or lagging at metaphase resulting in their loss or equal segregation. Supernumeraries are often eliminated from certain tissues at a particular stage of development. In *Haplopappus gracilis* they are of variable occurrence in shoot and are eliminated from root and contrarily in *Haplopappus spinolosum* they are contant in roots but variable in gernline. In *Sorghum purpureosericeum* they are retained only in reproductive organs.

(C) Impact on efficiency of plant

The impact of B-chromosomes on efficiency of a plant is anomalous, for example in *Secale cereale* increased sowing density and selection pressure reduced the survival of *Secale cereale* with B-chromosomes but increased efficiency in *Lolium perenne* plants with two B-chromosomes in comparison with the individuals deficient for them.Their absence or presence either increased or decreased the chaismata frequency and homoeologous pairing.In *Triticum* hybrids their high number generally have a deleterious effect on phenotype and fertility while a few supernumeraries caused delayed maturation and reduced fertility in *Anthoxanthum aristatum* (o¨stergren 1947).

2. Sex chromosomes in plants

In higher plants where male female flowers are borne on separate plants (dioecism) is a rare phenomenon. Generally in plants sex organs are produced in response to developmental factors and sex expression is highly susceptible to environmental conditions (Heslop-Harrison, 1975). In dioecious species, the sex determination is genetically controlled by alleles at a single gene locus. A cross between dioecious and monoecious forms of *Ecballium elaterium* suggested presence of 3 alleles at x locus,ie, (X^M-male), (X^+-monoecious) and (X^f–female) with relative dominance in that order.In plants normally males are heterogametic and females homogametic. In angiosperms only *Fragraria* (Westergaard, 1958) and *Potentilla fruticosa* (Grewal and Ellis, 1972) are reported to have heterogametic females. A considerable differentiation is reported to have occurred in chromosomes and hence their pairing ability reduced and they became morphologically distinct from each other as well as from other autosomes.

Cytologically identifiable chromosomes are known in some angiosperms like *Cannabis, Humulus, Spinacia* etc. At meiosis in sporophyte of Liverwort, Sphaerocarpus donellii, the X chromosome is large and Y is small. These x and y chromosomes segregate to half the spores, those with the X chromosome give rise to female gametophyte and others with Y to males. *In Silene album (Westergaard (1958)*,a diploid species. (2n=24), the female plants have XX chromosome and males have XY pair where Y is larger than X. Through induced polyploidy plants were produced with 2, 3 and 4 sets of autosomes and upto 5X and 2Y chromosomes. It was obsereved in general that presence of the Y chromosome gives rise to a male and its absence a female irrespective of number of X chromosome and sets of autosomes. The only exception are tetraploid plants with 4X and 1Y which are usually hermaphrodite. Chromosome fragmentation by X-ray produced plants with partly deleted Y chromosomes and study of these plants suggested that one part of Y chromosome contained anther development genes and other parts had genes suppressing the development of female characters. Thus in *Silene* sex is determined by balance between X and Y chromosomes without any role of autosomes, a ratio of 4X :1Y is essential even to produce an hermaphrodite. While in *Rumex acetosa* and related species a different situation exists where sex is determined by the balance between autosomes and x chromosomes and presence of y chromosome is of little importance. Plants with ratio of 1X:1A (where A = 6 autosomes) were female and plants with ratios of 1X :2 or 3 A were male; while 3A +2X , 4A+3X , 6A+4X were intermidiates with or without Y chromosome. These species also demonstrate how the sex chromosome mechanism can be modified by inclusion of one or more pairs of autosomes as a result of translocation. In *Rumex hastatulus* (Smith 1972) females with 2n=10=2A+XX and males with 2n=10=2A+XY, while in its another race females had 2n=8=2A+XX and males were with $2n=9=2A+XY_1 Y_2$. In race with $2n=9=2A+XY_1 Y_2$; $XY_1 Y_2$ forms a trivalent at meiosis and resulting in formation of X_1 or $Y_1 Y_2$ pollens. Cytological analysis of hybrids of this race showed that one of the autosome pair is homologous with an end of the X-chromosome and one of the Y-chromosome, which indicated occurrence of an interchange.Such interchanges between autosomes and X chromosome had the effect of bringing differentiation mechanism into the sex chromosome. Repeated translocations can include more autosomes as in *Humulus* where five chromosomes were are involved in sex determination.

The X and Y chromosome not only differ in size but they also differ in chromatin and in their replication as reported in mammals,where Y chromosome is often heterochromatic and one of the X of females becomes so during early embryogenesis.

3. Artificial chromosomes

Introduction of novel genes into plants and animals is accomplished through various methods such as microinjection , agrobacterium mediated transfer and transfection. However, these all methods depend upon DNA repair mechanisms of the target cell to insert DNA into chromosomes at a random location in the genome.However, such random insertion may potentially disrupt the function of other important genes. Artificial chromosomes do not require insertion of exogenous DNA into already existing chromosomes of the organism concerned. Moreover, genes located on the artificial chromosome are all linked and hence do not segregate independently and it is easy to transform these genes into other backgrounds also. Plant artficial chromosomes are chromosome based high capacity vectors (minichromosomes) for genetic manipulation. These are small and have no genes of their own and thus can be used as super vectors to accommodate and express a big amount of foreign gene complex in comparison to some other vectors. Secondly they are stable during both meiosis and mitosis and allow the resident genes to be expressed and transmitted faithfully from generations to generations.Thirdly they allow the addition, deletion and replacement of genes on them by either SSR systems or direct editing with recent genome editing technology.

Components of artificial chromosome and its synthesis

The analysis the large eukaryotic genomes has necessitated the development of high capacity vectors-artificial chromosomes or mini chromosomes. These are usually synthesized in small size with only essential components such as:

1. A centromere region;
2. Telomere region at each end;
3. Multiple origin of replication.

These three features are essential for basic functions of a chromosome. The centromeric region is required in order for the chromosome to become attached to the mitotic spindle during the cell division and to ensure proper segregation of daughter chromosomes.The telomeric regions are also essential and they are constructed in such a manner to ensure that the ends of DNA double helix are located in these regions. Although the molecular details of the processes involved at the telomeres at the tip of each chromosome are not fully understood but chromosomes without telomeres are not properly replicated and are eventually lost. Finally, multiple origins of replication are required in order to complete DNA replication in a reasonable period of time.

There are generally two methods for artificial chromosome construction: **de novo** assembly and **chromosomal truncations.**

De novo

The components of chromosome centromere repeats, telomeres and origins of replication are cloned and assembled in vitro. Next these desired contents of the artificial chromosome are transformed into a host which is capable of assembling the components (yeast or mammalian cell) into a functional chromosome. This approach is more difficult than truncated approach due to species incompatibility and heterochromatic nature of centromeric regions.

Chromosome truncation

Minichromosome construction by telomere mediated chromosomal truncation (TMCT) has been more successful which is based on discovery that transformation of telomere-containing sequences into genome will seed new telomeres at the site of integration (Farr et al; 1992 and Yu et al; 2006). Insertion of such telomeric sequences causes the generation of more telomeric sequences and eventual chromosomal truncation. The newly synthesized truncated chromosome can be manipulated through insertion of new genes for desired traits. The stability of these chromosomes during cell division has been demonstrated.

Artificial chromosomes are mainly used as cloning vectors that can accommodate very large segments of foreign DNA for producing recombinant DNA molecules. These chromosomes, resembling to a small chromosome, can be used in genomic study where entire genome can be managed in a number of clones. Two examples of artificial chromosomes (bacterial artificial chromosome and yeast artificial chromosome are being cited in the present context.

Bacterial artificial chromosome(BACs," backs")

Bacterial artificial chromosomes contain the origin of replication (f-factor) from a natural plasmid found in *E. coli*, a multiple cloning site, and one or more selectable markers (Fig.3). This particular vector can be used with blue-white colony screening method just like a plasmid.The selectable marker for this BAC is cam^R which encodes an enzyme, responsible for degradation of chloamphenicol,

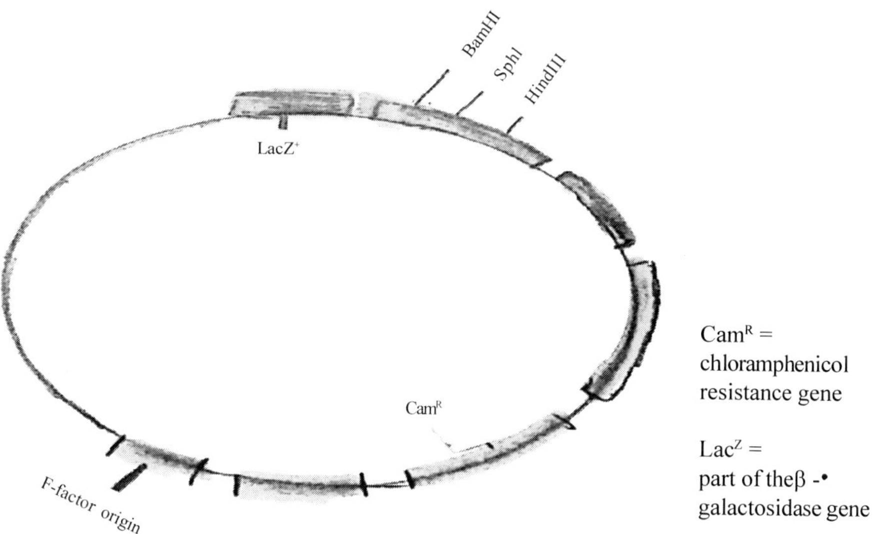

Fig. 3: Bacterial artificial chromosome (PBeloBAC11C) with selectable marker (cam^R) for chloramphrnicol resistance, a multiple cloning site in part of the lac^{Z+} gene, an origin derived from the F factor to limit the copy number of the BAC to one per *E. coli* cell.

and thus cells carrying this vector (with or without an insert) can grow in the presence of chloramphenicol while cells lacking this vector are unable to grow if chloramphencol is present. BACs can accommodate inserts upto size of 300kb and have the advantage of manipulation like giant bacterial plasmids. The one major difference between BACs and plasmids lies in the fact that transformed *E-coli*, due to F factor origin of replication, only one copy of the BAC is maintained per cell, while the origins of a plasmid cloning vectors drive multiple rounds of DNA replication to generate many copies of the plasmid in each cell. Since BACs do not undergo rearrangements in the cell like YACs, therefore, they are preferred vector for cloning large clones in physical mapping studies of genomics. The disadvantages of BACs are that AT – rich DNA fragments (high proportion of A and T nucleotides in DNA fragments) do not clone well.

Yeast artificial chromosome (YACs; "yaks")

These are also cloning vectors which enable artificial chromosomes to be made and replicated in yeast cells. YAC vectors can accommodate DNA fragments that are several hundred kilobase long and much longer than the fragments that can be cloned in a plasmid, cosmid & BAC vectors. Hence, YAC vectors are used to clone large DNA fragments. The main features of YAC vector are as:

1. Yeast telomeres (TEL) are of regularly repeating units (5'CCCA3') of about 100 bp in length. They preserve the ends of chromosomes by ensuring the completion of DNA replication.
2. Yeast centromeric and an autonomously replicating sequence (ars) enhances the stability of the construct especially with large inserts of DNA.
3. A selectable marker on each arm for detecting and maintaining the YAC in yeast, ie, trp1 and URA3 to enable transformed trp1 (tryptophan requiring) and URA3 (uracil requiring) mutant yeast to grow on medium lacking tryptophan and uracil, respectively.
4. An origin of replication sequence ARS (autonomously replicating sequence)- that allows vector to replicate in yeast cell.
5. An origin of replication (ori) that allows a circularization of the empty vector to replicate in E.coli and a selectable marker such as amp^R that functions in *E.coli*.
6. A cloning region that contains one or more restriction sites; the restriction enzyme cutting into an region should not have any other sites in the YAC to be used for inserting foreign DNA.

Empty YAC vectors – ones that have yet to contain a DNA insert –are propagated in *E. coli* as circular plasmids; in this form two telomeres are end to end. This propagation step makes use of the bacterial origin of replication and bacterial selectable marker. The bacterial and eukaryotic origins of replication are not functionally similar, which means that the yeast ARS sequence will not work in a bacterial cell, just as the bacterial ori sequence in a yeast cell. In addition, bacterial and eukaryotic promoters are different since that bacterial DNA polymerase can not transcribe the yeast TRP1 and URA3genes, so those selectable markers will function only in yeast, not in bacteria. Likewise yeast RNA polymerase II is unable to transcribe the amp ^{R} gene.

For cloning purpose, a circular YAC is cut with one restriction enzyme that generates cuts in the multiple cloning sites and with another restriction enzyme between two TELs. In this way the left and right arms are produced (Fig. 4). High molecular weight DNA, cleaved with the same restriction enzyme used to cut the YAC multiple cloning site, is ligated to two arms and the recombinant molecules are transformed into yeast. By selecting for both TRP1 and URA3, it can be ascertained that the transformants have both the left and right arms

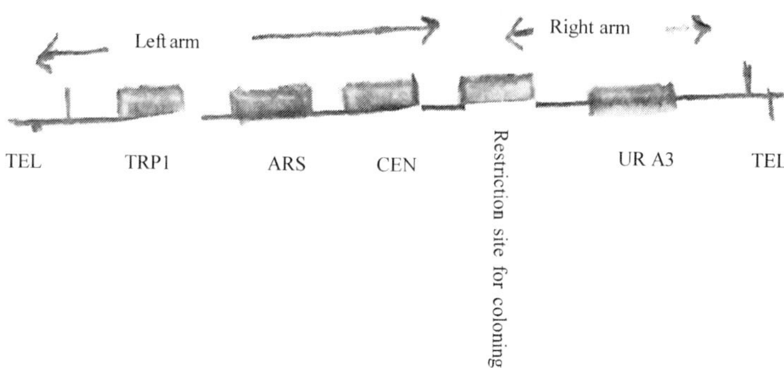

Fig. 4: A yeast artificial chromosome (YAC) (Linear form) contains a yeast telomere (TEL) at each end, a yeast centromere sequence (CEN), a yeast selectable marker (for each arm (TRP1 and URA3) and restriction site for cloning.

References

1. Belling, J. (1921) Amer. Nat. 55: 573-574.
2. Farr C.J., Miilena S., Thomson, E.J., Goodfellow, P.N. and Howard, J. (1992). Nature Genetics 2(4): 275.
3. Feulgen, R. and Rossenbeck, H. (1924), Hoppe-Seyl Zeit; 135: 203-248.
4. Grewa L, M.S. and Ellis, J.R. (1972). Heredity, 29: 359-362.
5. Heslop-Harrison, J. (1957). Biol. Rev; 32: 38-90.
6. Lima-de-Faria, A. (1949), Hereditas (Lund), 35: 77-85.
7. Mc Clintock, B. (1929). Science 69: 629-630.

8. McClintock, B. (1930). Proc. Nat. Acad. Sci. (Wash). 16: 791-796.
9. Östergren, G. (1947). Hereditas (lund), 33: 261-296.
10. Sharp, L. (1929). Bot. Gaz. 88: 349-382.
11. Smith, B.W. (1972). Chromosomes Today, 2: 172-182.
12. Taylor, W.R. (1925). Amer. Jour. Bot. 12: 238-244.
13. Westergaard, M. (1958), Adv. Genet.; 9: 217-281.
14. Yu W., Lamb J.C.; Han F. and Birchler, J.A. (2006) Proc. Natl. Acad. Sci. U.S.A. 14; 103(46): 17331.

2

Chromosomal Theory of Inheritance

The relationship between gene and chromosome was suspected at the time of rediscovery of Mendelian Laws. During that period Montgomery (1901) and Walter Sutton (1902) working on grasshoppers showed that chromosomes occur in distinct pairs, often of recognizable shape and size, and that synapsis involves the union of maternal and paternal chromosomes, while Winiwarter (1901) concluded from his studies in rabbit ovaries that bivalents in the first meiotic division resulted from the chromosome pairing side by side and not end to end as believed by some other workers of that time . Theodor Bovery (1902) showed from his work on polyspermy in the fertilization of sea urchin eggs that the chromosomes of an individual were not equivalent (genetic information) to one another and hence a full chromosomal complement is essential for normal development of cell. Correns (1902) and Cannon (1902) pointed out close parallelism between Mendelian segregation and chromosome reduction during meiosis and concluded that genes are located on chromosomes but they, like de Vries, supposed that maternal and paternal chromosomes move to opposite poles during meiosis. In the same year Guyer established an understanding that random assortment between different pairs of chromosomes would give the independent assortment of genes required by Mendel. Ultimately Sutton (1903) brought together the ideas from cytology and genetics to clearly show the role of chromosomes in heredity and hence to establish the field of cytogenetics. Boveri also through his work, advanced many of the same ideas and thus the hypothesis correlating genes and chromosome transmission was established, which is known as the **"Sutton-Boveri Hypothesis or chromosome theory of inheritance".**

The salient features of the hypothesis are as follows

1. In somatic cells there are two similar groups of chromosomes, one of paternal and other of maternal origin, This occurrence of chromosomes in homologus pairs parallels the occurrence of genes in pairs (alleles).
2. The chromosomes retain a morphological individuality throughout the various cell divisions; genes also show the similar continuity.

3. During meiosis homologous pairs of chromosomes are brought together and the members of each pair segregate into germ cells independently of the members of other pairs; Mendelian genes too segregate independently at some time prior to the gamete formation.
4. Each chromosome, or chromosome pair, has a definite role in life and development of the individuals

Sutton (1910) alongwith establishing the relationship between gene and chromosome also recognized that there must be non-independent assortment of some genes too (linkage) otherwise the number of distinct characters would not exceed the number of chromosomes.

The association of a particular inherited character with a particular chromosome was established between 1901 to 1906 by McClung, Stevens, Wilson and others who showed that in Hemiptera and orthoptera, females have one more chromosome (X chromosome) than the males This so called x chromosome occurs in all eggs but in only 50% of sperm so that half of the resultant zygotes are XX (female) and half are XO (males). The presence of small y chromosome, partially homologous with X chromosome, in males of beetles, insects, mammals and other groups confirmed the same pattern –that the sex chromosome of the male gametes determines the sex of progeny. While the discovery of the reverse situation of female heterozygosity in birds and Lepidoptera also confirmed the importance of chromosomes in sex determination.

The association of a particular gene with a particular chromosome was demonstrated by T. H. Morgan (1910) who showed that the inheritance of recessive allele (w) for white eye in Drosophila paralleled that of the x chromosome (Fig. 5A and 5B). However, conclusive evidence that the white locus was situated on the x chromosome was provided by Bridges (1916) who reported that sometimes a cross between a white eyed female and a red eyed male produced a white eyed female or red eyed male among the F1 progeny. This was found to be due to non-separation or non disjunction of x chromosome at meiosis in female resulting into eggs with either two or no x chromosome. (Fig. 5C) The xxy constitution of the white eyed females was confirmed cytologically. Thus Bridges found the correlation between chromosomes and genes through the first critical evidence that genes are located on chromosomes.

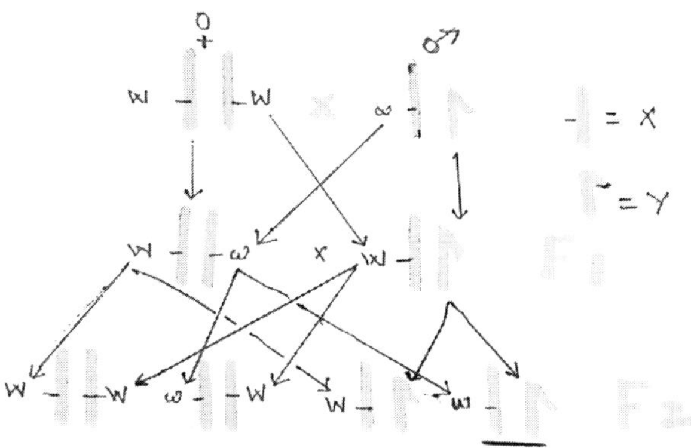

Fig. 5A: Inheritance of red (W) and white (w) alleles for eye colour in Drosophila

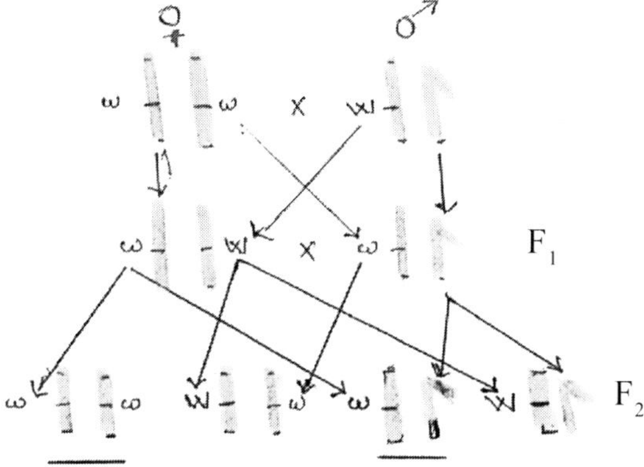

Fig. 5B: (Reciprocal of 5A) shows parallel transmission of W allele and x chromosome (♀) in reciprocal cross. Underlined progenies are white eyed (Morgan 1910)

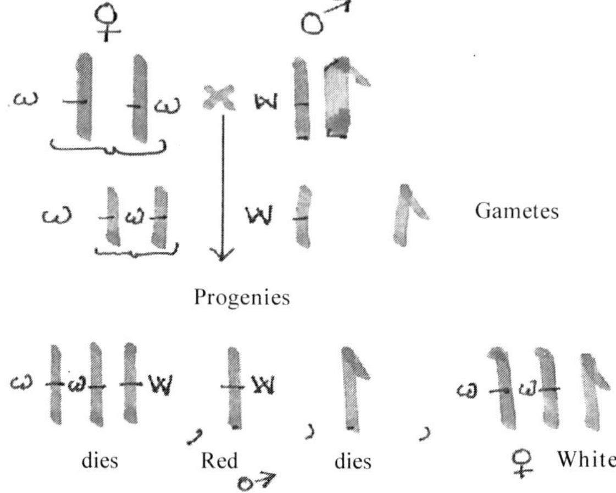

Fig. 5C: A cross between white eye female (ww) with non-disjunction of X (♀) chromosomes and red eye male in Drosophila producing exceptional types of progenies (Bridges 1916)

Linkage, crossing over and chromosome

As earlier mentioned that Sutton pointed out that if there were more gene loci than chromosomes, a fact since abundantly demonstrated in all plants and animals, then chromosome theory of inheritance would not permit the Mendelian law of independent segregation to apply to genes located on the same chromosome. The data of Bateson and Punnet (1905-08) on Sweet Peas provided the genetical evidence of linkage and Sturtevant (1916) demonstrated the linear arrangement of genes on chromosome and the use of 3-point testcross for mapping the loci. Finally Stern (1931), Brink and Cooper (1935) and Creighton and Mc Clintock (1931) resolved the relationship between chromosome, genes and phenomenon of independent assortment. They demonstrated genetic exchanges have had cytological exchange too between the two members of a chromosome pair, a heteromorphic one, marked cytologically at two points and analyzing the genetic crossovers cytologically in drosophila and corn, respectively. The work of Creighton and McClintock (1931) elucidated clearly the relationship between cytological and genetic recombination in maize (Fig.6). One experimental plant was heterozygous for a chromosome with terminal knob and a long extra segment (—) and a normal chromosome (N) without knob and for alleles determining colored (C) or colorless (c) and waxy (wx) or nonwaxy (-starchy) endorsperm. Mc Clintock and Creighton followed the knob on a specific chromosome in maize gametes and correlated cytological marker with specific phenotypes. For example, in the parental genotype knob is located on a chromosome having a extra segment of a chromosome. However, in test cross progenies the normal chromosome also had knobbed feature or chromosome with extra segment without knob. It is only possible when genetic recombination has been accompanied by physical exchange between a homologous pair. The same was true in case of genetic markers as coloured (C) or colorless (c) and non waxy (Wx) or waxy (wx). The experimental details of Creighton and Mc Clintock experiment have been presented in Figure 6. The experimental results established the relationship between cytological and genetical recombination and how linked genes (genes on a chromosome) undergo independent assortment.

The cell cycle and cell division

The cell cycle is the sequence of events between successive cell divisions.During the cell cycle many different processes must be coordinated, so that cell size and its DNA content remain constant. Some of the processes of cell cycle occur continuously (e.g .cell growth) and some discontinuously (e.g. cell division). Cell division must be coordinated with growth and DNA replication. The cell cycle comprised of a nuclear or chromosomal cycle (DNA replication and partitioning) and a cytoplasmic or cell division cycle (doubling and division of cytoplasmic components.

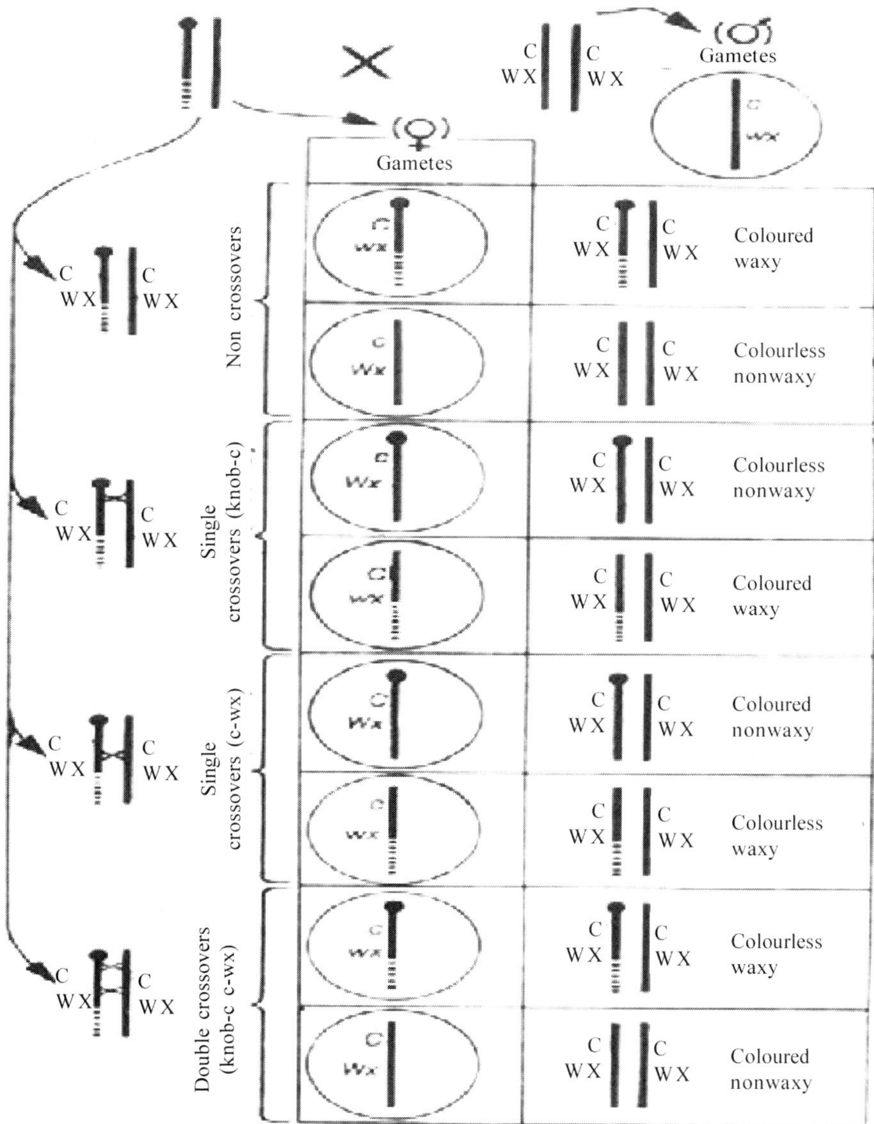

Fig. 6: Test cross in maize demonstrating correlation between recombination and crossing over. One plant is heterozygous for a chromosome with terminal knob and a long extra segment (——) and a normal chromosome (N) without knob but having alleles determining colored(C) or colorless (c) and waxy (wx) or nonwaxy (-starchy) endorsperm.(Creighton and Mc Clintock 1931) Migration of knob from chromosome having extra segment (—) to chromosome without extra segment and genetic markers establishes that genetic exchanges are accompanied by cytological exchanges also.

The cell cycle is controlled by protein kinases and its transition involves positive feedback loops which cause sudden bursts of kinase activity, allowing switches in the states of phosphorylation of batteries of effector proteins. Cell cycle check points are regulatory systems which inhibit those kinases if internal or external environment is unsuitable. The alteration of DNA replication and mitosis is controlled by negative feedback- mitosis is inhibited by unfinished DNA replication, and DNA replication is prevented during mitosis by the phosphorylation and inactivation of a protein required for replication. The cell cycle is the result of complex network of information, in which kinases are controlled by the integration of multiple positive and negative signals.

For cytogenetical studies the main emphasis is given on the behavior of chromosomes during cell division, since chromonemata are present throughout the cell cycle and maintain the genetic continuity between cell and organism-generations.The non dividing cell has often been described as having a resting stage nucleus, because of the invisibility of the chromosomes with usual stains. But as a matter of fact interphase is not a resting stage, since it is metabolically a very active stage in controlling the metabolic activity of the cell and replicating DNA prior to the next phase of chromosomal division. Based upon the chromosomal activity cells can be categorized as **autosynthetic** (engaged in active proliferation), **heterosynthetic** (differentiating cells) and **non-synthetic** cells (completed differentiation and with low metabolic activity). In present context only autosynthetic cells will be considered which are actively proliferating and engaged in synthesis of DNA, a prominent feature of interphase, prior to the next mitotic division. It is customary to recognize four stages of cell cycle, ie, G1, S, G2 and M (Fig. 7).

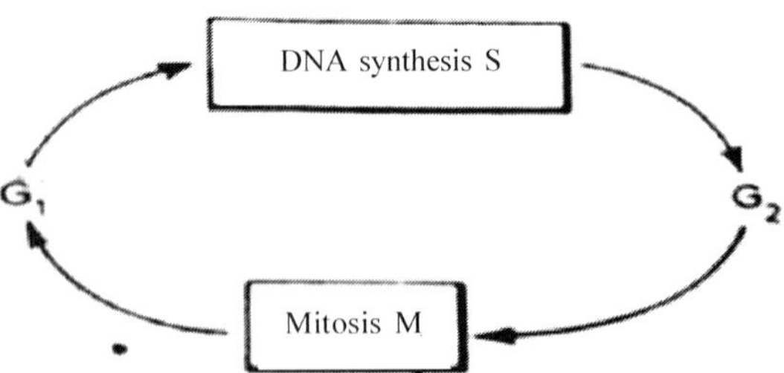

Fig. 7: Generalized cell cycle of a dividing cell.

The mitotic (M), and DNA synthesis phases (S) are separated by presynthetic (G_1) and post synthetic (G_2) periods. The autosynthetic cell cycle lasts for approximately 17-32 hours, of which mitosis occupies about 1.5-4 hours (John and Lewis, 1969). Out of the four principal mitotic phases telopase is usually the longest while prophase, metaphase, and anaphasae being progressively shorter(Mazia, 1961). The duration of complete cell cycle is affected by general metabolic features such as respiration and nuclear DNA content.The duration of S phase is closely correlated with that of the cell cycle and increases linearly with DNA value. Variation in the duration of cell cycle between taxa have been demonstrated and it was found to be longer in dicotyledons than in monocotyledons because of much longer G_1 period in the former which have highly coiled metaphase chromosomes requiring longer period to uncoil during G_1 prior to the initiation of synthesis. However, there is still a good deal of conflicting evidence on these points and probably some other parameters are also involved as longer cell cycles and S phases in perennial than annual members of several genera of compositae were not positively correlated with the nuclear DNA content (Nagl and Ehrendorfer, 1974). Autosynthetic cells can give rise to heterosynthetic and non synthetic cells or, of major importance they can give rise to sexual, germline cells which eventually lead to meiotic cycle. Many of the features of mitotic cell cycle are shared by meiotic cells but the later have some specific properties as observation on meiotic chromosomes have shown that it generally takes much longer time than mitosis, lasting about 6 days in *Tulbaghia*, 24 days in *Lilium* and *Trillium.* The length of cycle is greatly modified by various environmental factors, which also seem to affect other components of plant growth.

In general it seems that meiosis begins earlier in relation to the inception of the cell than does mitosis and hence having no obvious G_2 period. Although the onset of the long meiotic prophase is difficult to determine precisely but S phase seems to occur largely during late interphase, when 99% of DNA synthesis occurs (Ito et.al, 1967) or at late leptotene. Small amount of DNA are also synthesized at zygotene and pachytene, but these appear to be associated with chromosome pairing. In addition to the phases of DNA replication, both meiotic and mitotic cell cycles show striking periodicity in synthesis of RNA and various classes of proteins. Their occurrence is profoundly important in relation to the metabolic activity of the cell and the dissemination of the genetic information conserved by chromosome. Chromosomal histone is synthesized along with DNA and during this time synthesis of nuclear RNA is low or absent.

Chromosomal replication

During meiosis and mitosis the replication of chromosomes, by prophase in the former and by diplotene in the latter, gives visible evidence of their suitability for providing the physical continuity throughout the development of the individual and between generations required of the carriers of genetic information. With the realization of genetic importance of DNA and the discovery that it doubles in amount prior to meiosis and mitosis, it became apparent that, in some way, that two sister chromatids (replicated form) of a chromosome (unreplicated form) are visible result of DNA replication which has occurred during interphase. Moreover, sister chromatids, can not be recognized as such in the interphase, ie, in the G_1, S, or G_2 phase nucleus. It can only be observed when the nucleus is preparing to divide either by mitosis or by meiosis. Furthermore, there are obvious parallels between chromosomal replication and semi- conservative mode of replication of DNA (one old and one new nucleotide chain) postulated by Watson and Crick, 1953) and confirmed by (Taylor *et al.*, 1957) in Vicia.

The Watson and Crick model showed how DNA could replicate progrssively from one end of the molecule like the opening of a zip-fastener and there are now abundant data from prokaryotes to confirm that the genophore (Bacterial chromosome) does indeed replicate progressively in one or two directions from a single initiation point. In case of eukaryotic chromosome there is considerable evidence from autoradiographic studies to suggest that DNA synthesis is initiated at a number of different points. In *Scilla campanulata* interchromosomal differences in DNA replication were directly proportional to the chromosomal length, and replication near the centromere was completed earlier than the more distal regions (Evans and Rees, 1969) but in *Phaseolus* chromosomes replication begins simultaneously at several sites in the euchromatin, followed by the heterochromatin, except that near the centromere which replicates last (Brady and Clutter, 1974). These chromosomes also demonstrate the close control over replication but still there is good deal to be learned about the details of chromosome replication, particularly regard to the non-DNA components of the chromosomes. The doubling of DNA observed at mitotic and meiotic prophase depends upon the process of spiralization and condensation. The process of condensation, chromatids become small and deeply staining bodies (chromosome). During mitotic division each sister chromatid remains connected to a region known as centromere and effectively spilt into two, with one sister chromatid and its centromere half going to one daughter nucleus and the other chromatid with its centromere half to the second daughter nucleus. The separated chromatids are now called daughter chromosomes (Fig. 8).

Cell division (Mitosis and Meiosis)

'Mitotic Division is unrelated to DNA and chromosome replication, since these events occur during the synthetic phase (s) of interphase. Mitosis is an essential phenomenon for chromosomal duplication since it ensures that each daughter cell gets one sister chromatid of each chromatid pair of a chromosome and therefore a complete set of chromosomes in the resultant cells after mitotic division. Moreover, mitosis does co-ordinate DNA replication so that cell size and DNA content also remain constant during subsequent plant growth.The four mitotic stages –prophase, metaphase, anaphase and telophase are in fact arbitrarily defined, since mitosis occurs as a continuous process (Fig. 9).

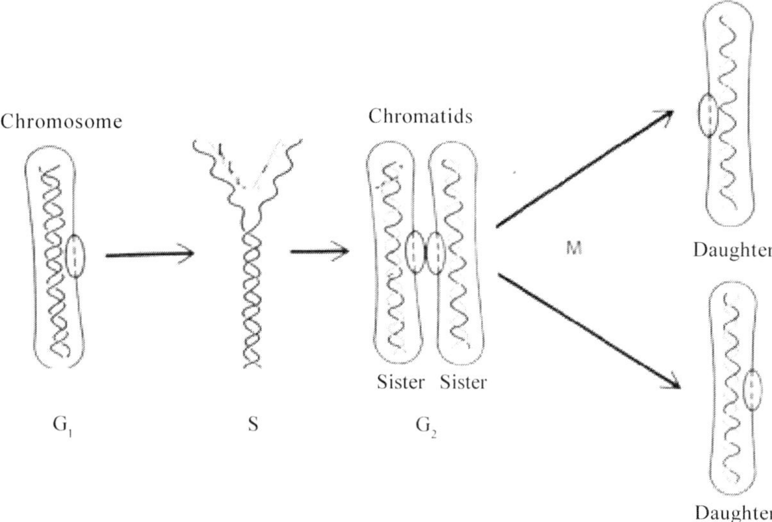

Fig. 8: The mitotic replication and separation of sister chromatids of a chromosome. The centromere is represented by an oval structure

Prophase: Prophase is heralded by the onset of chromosome coiling and condensing. As these processes progress, two sister chromatids of a chromosome can often be distinguished.The other notable feature of the prophase is that the nucleolus, a site of ribosomal RNA (rRNA) synthesis, becomes undetectable under light microscope and its component particles disperse throughout the nucleus (Fig 9a and 9b)

Metaphase: Metaphase begins with the breakdown of nuclear envelope and release of chromosome and nucleoplasm, filling the greater part of the cell. The other diagnostic feature of the metaphase is development of spindle apparatus.Under electron microscope each spindle, apporoximately 250 A^0 in diameter, is seen to be composed of numerous long fibres known as microtubules.Some of these microtubules attach to the kinetochore of chromosome. A kinetochre of a chromosome at metaphase can be seen exhibiting two faces, one pointing towards one pole and the other toward the other pole. When all the chromosomes are thus attached to the spindle they are guided to metaphase plate held by tension between opposing poles of spindle being attached to the sister kinetochore.

Anaphase: Anaphase begins with the division of centromeres or more precisely of the kinetochore. Paired kinetochores separate towards poles as kinetochore microtubules shorten and poles also move apart as polar microtubules repel each other. Once the chromatids have separated they are considered as daughter chromosomes or simply chromosomes of new resultant cell. The alignment of chromosome along the metaphase plate and their movement to poles at anaphase constitutes one of the most dramatic activities exhibited by eukaryotic cells.It has been proposed that certain contractile proteins participate in mitotic chromosomal movement.

Telophase: Chromosomes reach pole and nuclear envelope reforms around each set of daughter chromosomes along with nucleolus. The chromatin decondenses, spindle microtubules disappear and nucleus gradually assumes an interphase morphology.The telophase is followed by cell division or cytokinesis and thus producing two genetically identical cells.

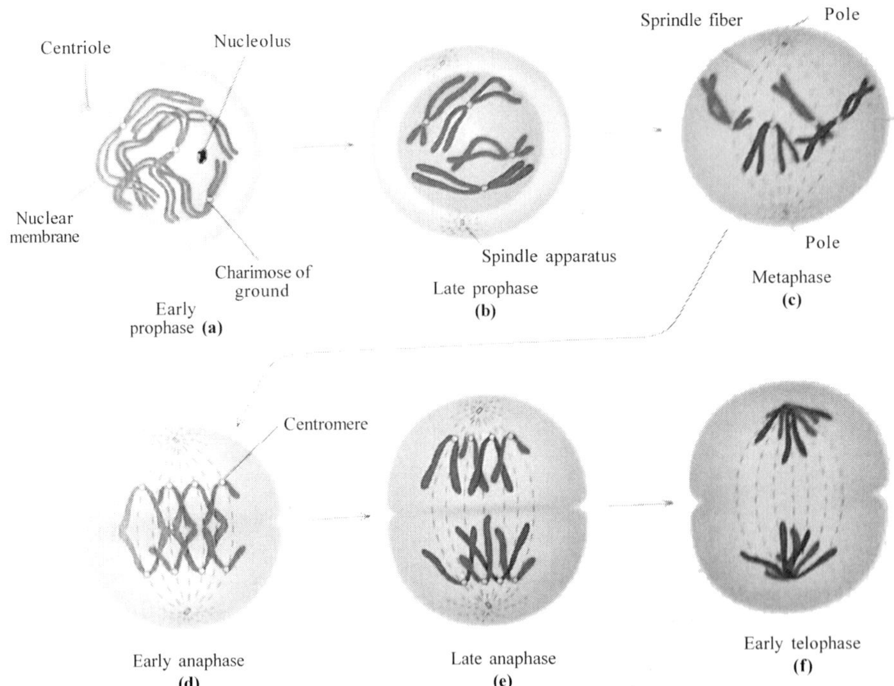

Fig. 9: Hypothetical mitotic stages with 2 pairs of chromosome

Meiosis

In meiosis chromosome doubling does not occur like mitosis and its S phase occurs well before its begining. For meiotic division a diploid cell enters both meiosis and mitosis with 4C amount of DNA. A basic difference between meiosis and mitosis, however, is that meiotic DNA replication is followed by two nuclear divisions in succession rather than one . It should be obvious that a diploid cell DNA (2C) replicates to become (4C) and then to undergo two meiotic divisions, the first division produces two cells (dyad)with a (2C) amount of DNA, while second division results in four cells(quartet) that are each haploid or C in the amount of DNA they contain. In other words, the net effect of meiosis is to reduce the somatic chromosome number in a gamete by half (2n to n) which is a prominent feature of meiotic division.

The diploid cells contains two complete set of chromosomes, ie, paternal and maternal. The meiotic reduction (2n to n) must occur in such a way that each haploid product of the meiosis is allotted one complete set of chromosomes (any one chromosome out of each homologous pair containing all genetic information pertaining to that species). The guarantee of such allotment of chromosomes, is **the second important feature** of the meiotic division.

A third key feature of the meiosis is that at a particular meiotic stage in eukaryotic development new gene combinations are generated in numerous combinations through orientation of homologous pairs at metaphase I and genetic exchange between homologous chromosome pairs during meiosis. Fig. 10 illustrates how variability can be generated during meiosis through random assortment of chromosomes to different poles. An offspring (AA' BB') from parents (AABB) and A'A'B'B') will produce four different types of gametes during meiosis, two of which are parental (AB and A'B') and two (A'B and AB') are new combinations produced by change of orientation of chromosomes. Secondly, maternally and paternally derived homologous chromosomes frequently take part in genetic exchange (crossing over) during meiotic prophase. And thus produced a haploid product may contain some maternally derived and some paternally derived alongwith some recombinant chromosomes as members of its complete chromosome set. The recombinant chromosomes (post recombination) contain mixed genetic information from chromosomes of both parents.

Meiotic division is comparatively lengthy than mitotic division and it usually takes days or weeks for completion rather than hours. Its first stage, prophase I, is complex and comprised of sub-stages: leptotene, zygotene, pachytene diplotene and diakinesis. Such division into sub-stages is arbitrary and all, of course, flow from one to the next. Prophase I is followed by metaphase I, anaphase I and telophase I and these are followed by prophase II, metaphase II, anaphase II and telophase II (quartet). A short interphase usually separates meiosis I and meiosis II in plants. Leptotene marks the end of pre-meiotic interphase.The literal meaning of leptotene is "slender thread" since during this stage chromosome look like threads with many beads like structures (chromomeres) along with their length Although replication of DNA has already occurred during S phase but at this stage it is difficult to observe any morphological duplication and generally chromosomes appear as single and individual structure until pachytene.

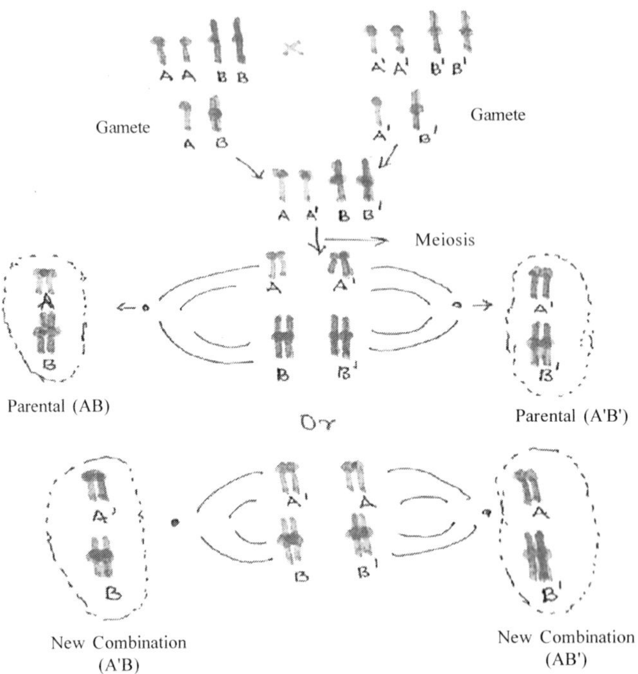

Fig. 10: Depicts the generation of variability through orientation of chromosomes during meiosis.. The F_1 (AA' BB') between AABB and A'A'B'B' produces two parental (AB) and A'B') and two new gametic combinations(A'B) ana (AB')

Zygotene: Zygotene (from the Greek zygon, a yoke) is defined as the stage in which homologous sets of sister chromatids complete their side to side alignment, an association called **synapsis.** It is highly specific and occurs between all homologous chromosome sections even present on different non homologous chromosomes (translocation or interchange). With the light microscope, a zygotene nucleus can be distinguished in two ways: as in many organisms the association of telomeres with the nuclear envelope causes a "bouquet" arrangement of the chromosome strands and in other organisms zygotene chromosome strands appear thicker than they do in leptotene. This thickness comes partly from chromosome condensation and partly from intimate intermeshing of sister chromatids with one another so that it becomes impossible to distinguish them as individual entities. Each synapsed set of homologues, which in fact includes four chromatids, appear to be composed of only two chromosomes **(bivalent).**

Pachytene: Pachytene is the stage of progressive shortening and coiling of the chromosomes. At pachytene non sister chromatids of a bivalent undergo a series of exchanges of genetic material. Such exchanges can be detected by cytological means and it signifies the genetic mechanism of crossing over or recombination.

At this stage a structure called **synaptonemal complex** can be observed between synapsed chromosomes after osmium-formaldehyde or glutaraldehyde fixation, by electron microscopic studies of meiotic prophase in a variety of animals and plants as a constant feature of eukaryotic meiosis. It appears in association with leptotene chromosomes and disappears by diplotene. The complex well developed at pachytene consisted of three parts: two intensely staining lateral elements about 45nm wide separated by a distance of about 100nm through which runs a central element; most of the chromosomal chromatin lies outside the complex, flanking the lateral elements (Fig. 11). The establishment of synaptonemal complex is crucial genetic event because these complexes mediate the meiotic exchange of genetic information known as **crossing over or recombination**, between homologous chromosomes.

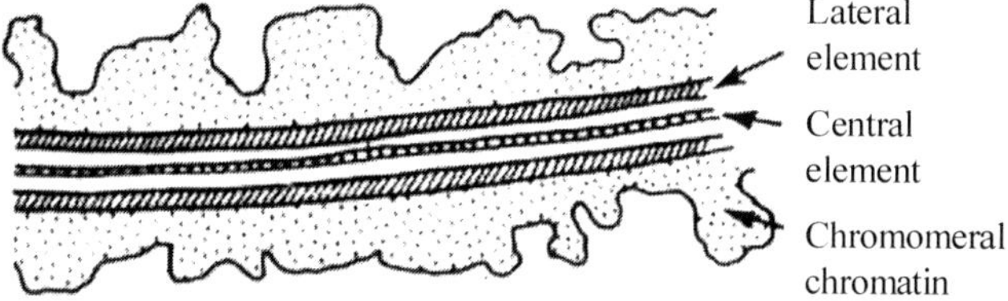

Fig. 11: Components of synaptonemal complex as seen under electron microscope (La Cour and Wells, 1974)

Distinctive DNA-protein complexes are produced at zygotene in *Lilium* (Hecht and Stern, 1971) and it has been shown by Hott *et al.*, 1968) that when their synthesis is blocked in Lilium microsporocytes, chromosome pairing and chiasma formation are prevented. Most of the chromatin of the bivalent is held outside the lateral elements of the synaptonemal complex, which suggest that only a small part of the total chromosomal content of nucleoprotein is involved in synapsis and chiasma formation (La cour and Wells, 1974). The synaptpnema complex, though necessary for synapsis, does not always result in chiasma formation in triploid *Allium.* (Stack, 1973) inspite of synapsis and complete development ofsynaptonemal complex.

Diplotene. During diplotene, the four chromatids of a bivalent under go repulsion. however, two kinds of constraints appear to prevent the repulsing chromosomes from separating completely, first the sister chromatids continue to be held together by centromeres. Second, non sister chromatids in a bivalent are usually held together at one or several positions along the length of chromosomes. These apparent contacts are called chiasmata (singular-chiasma). Chiasma means a cross and chromatids connected by a single chiasma typically assume a cross like configuration when viewed under light microscope.Two, three or all four chromatids in a bivalent may participate in chiasma formation with nonsister chromaqtids and a given chromatid may form more than one chiasma. On the basis of the involvement of chromatids in recombination, the crossing over can be termed as two strand, three strand and four strand one with consequences shown in the Figure 12. The occurrence of chiasmata is the normal feature of meiosis and there is atleast one chiasma per bivalent.

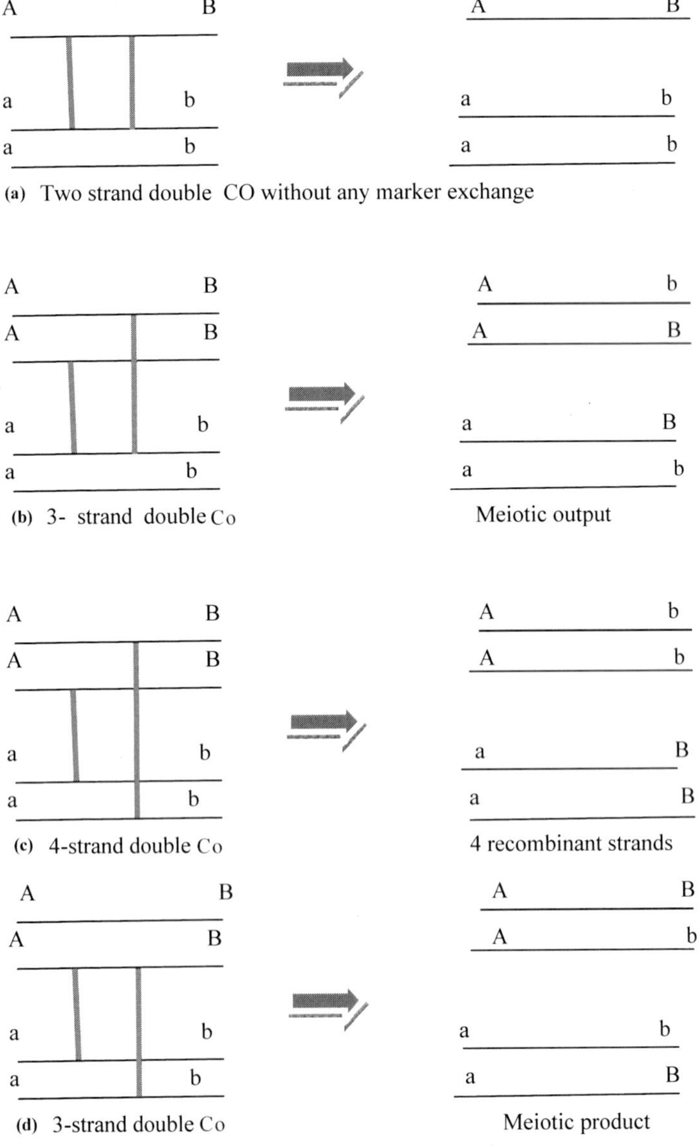

Fig. 12: a,b,c and d show 2 strand, 3 strand and 4 strand double crossing over and its consequences

Diakinesis: Throughout the prophase I chromosomes have continued to coil and by diakinesis they are maximally condensed. During this stage chiasmata frequently go through a process known as **terminalization** (moving towards the end of bivalent) . The bivalents are thus characteristically interconnected by one or two chiasmata as they appear at metaphase I. The nucleolus either disappears or detaches from its associated chromosome. In the end of this stage or in the early part of metaphase, the nuclear membrane dissolves and the bivalents attach themselves with rapidly formed spindle.

Metaphase I: Metaphase of the first meiotic division is characterized by spindle formation, as in mitosis, but the two divisions are otherwise different. Each bivalent includes two distinct centromere, each of which holding their respective chromatids.Two centromeres align themselves on either side of metaphase plate. For meioses, more than one bivalent involves and each involved bivalent posseses one maternally derived and one paternally deriveed centromere. These maternal and paternal centromeres have equal probability to align themselves either of the upside or lower side of metaphase plate (Fig.10). In other words, the maternal-paternal arrangement of a given set of homologous centromeres with respect to the metaphase plate is totally independent of the arrangement assumed by all other sets and all possible combinations of arrangements thus occur with equal frequency when large numbers of metaphase I cells are considered. This genetic fact is the basic cause of Mendelian independent assortment.

Anaphase I: Although already duplicated each chromosome, still maintains single functional centromere for both of its chromatids. The separation of one homologous chromosome from another in anaphase I towards opposite poles, therefore, results with a single centromere (which may already be spilt). As the centromeres are pulled apart towards opposite poles, the chiasmata further slips of the ends of chromosomes (terminalization). For pole ward movement the chromosomes are only attached with centromere.

Usually the meiotic division is considered as reduction division since two homologous chromosomes of a bivalent are separated into two daughter cells. However, this concept is modified in accordance with the exchange of chromosomal material manifested by chiasmata Between a pair of homologous chromosomes, if no exchange (crossing over) occurs in meiosis the separation is purely reductional, since chromosomes remain intact in their original form. But if genetic exchange (crossing over) occured each separating chromosome carries part of its homologue, resulting in an equal (equational) division of the exchanged chromosome material to both daughter cells. Thus each chromosome consists of genetic material from both chromosomes of a homologous pair or from both parents. (Fig 13)

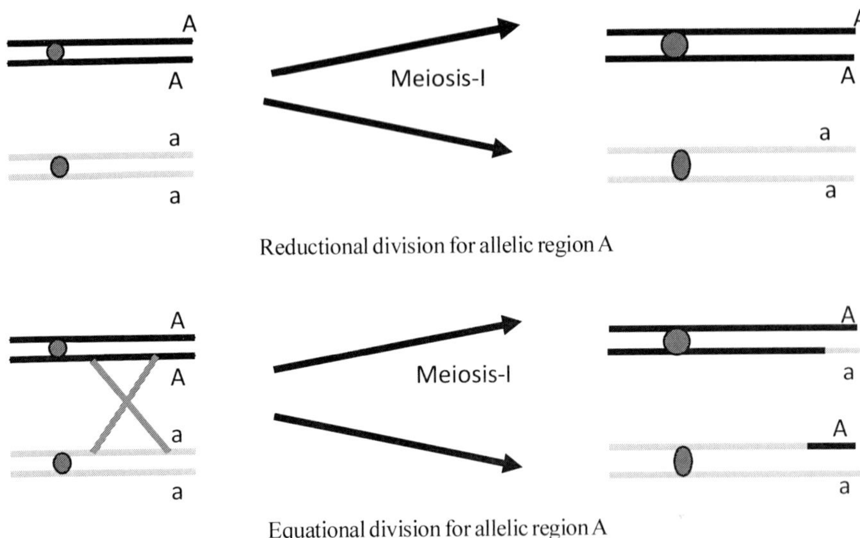

Fig.13: Mechanism of reductional and equational segregation

Gametes formed from such daughter cells may, therefore, diversify the genetic base of progeny of next generation..

Telophase I: Once the separated chromosomes from bivalents reach to poles of spindle, a nuclear membrane is formed around two separated sets of homologous chromosomes or chromosomes may enter directly into second miotic division. Similarly mechanical division of the cell (cytokinesis) may occur or may be postponed until the formation of quartet at the end of meiosis II.

Meiosis II: (prophase, metaphase, anaphase and telophase II) : Cells that bypass a telophase I, enter the meiosis II directly without experiencing prophase II and the second meiotic division is morphologically indistinguishable from a mitotic division. Centromeres connecting pairs of chromatids move to a metaphase plate and divide into two halves and move to opposite poles like mitotic division. At the completion of telophase II, therefore, four haploid cells or a tetrad/ quartet is derived from each original diploid cell (Fig15).

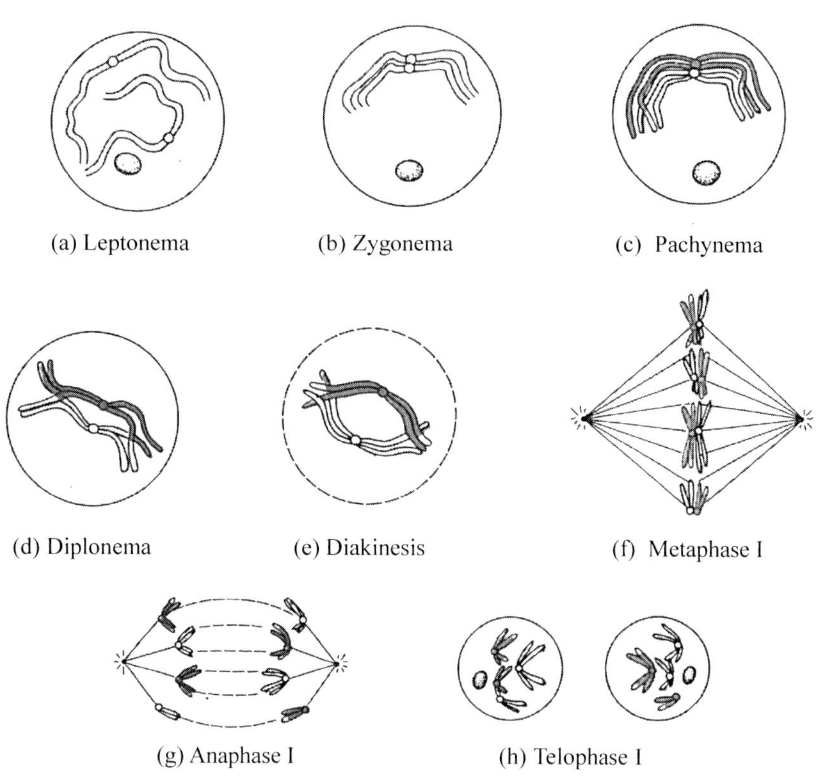

Fig. 14: Diagramatic representation of various meiotic stages of meiosis I.

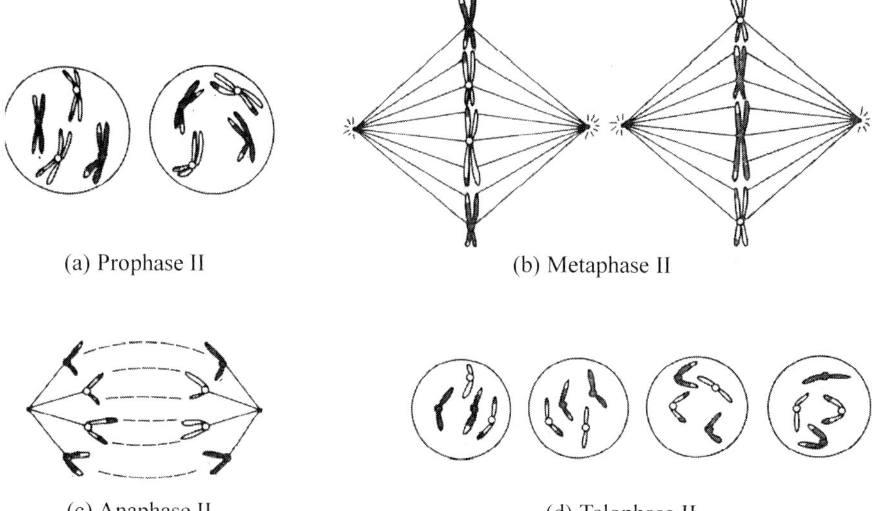

Fig. 15: Various stages of meiosis II.

Mitosis	Meiosis
1. An equational division which separates sister chromatids.	1. The first stage is reductional division which separates homologous chromosomes at first anaphase; sister chromatids separate in an equational division at second anaphase
2. One division per cycle, i.e. one cytoplasmic division (cytokinesis) per equational chromosomal division	2. Two divisions per cycle, i.e. two cytoplasmic divisions, one following reductional chromosomal division and one following equational chromosomal division.
3. Chromosomes fail to synapse; no chiasmata formed; genetic exchange between homologous chromosomes does not occur.	3. Chromosomes synapse and form chiasmata; genetic exchange occurs between homologues.
4. Two products (daughter cells) produced per cycle.	4. Four cellular products (tetrad) (gametes) produced per cycle.
5. Genetic content of mitotic products are identical.	5. Genetic content of meiotic products different; centromere may be replicas of either maternal or paternal centromeres in varying combinations.
6. Chromosome number of daughter cells is the same as that of mother cell.	6. Chromosome number of meiotic products is half that of mother cell.
7. Mitotic products are usually capable of undergoing additional mitotic divisions.	7. Meiotic products cannot undergo another meiotic cycle although they may undergo mitotic division.
8. Normally occurs in most all somatic cells.	8. Occurs only in specialized cells of the germline.
9. Begins at the zygote stage and continues through the life of the organism.	9. Occurs only after a higher organism has begun to mature; occurs in the zygote of many algae and fungi.

Deviations from normal meiosis

A Meiotic drive

Meiotic drive refers to any process which alters the normal process of meiosis with consequence that a heterozygote for two genetic alternatives produces an effective gametic pool with an excess of one type whch causes some alleles to be over represented in the gametes formed during meiosis and thus leading to violation of Mendelian expectations since under nomal meiosis each chromosome (and gene) is consistently present in about half of the meiotic products. The process of meiotic drive increases the frequency of certain alleles by the meiotic mechanism even when opposed by selection (Fig.16). The most fully explained example of meiotic drive is the preferential segregation of abnormal chromosome 10 (K10) with a large terminal knob in maize. The large terminal knob, and associated genes, on K10 are recovered normally through the pollen but in about 70 % of eggs from K10/k10 heterozygotes. Other chromosomes which are heterozygous for knob will do likewise, irrespective of presence of heterozygous or homozygous

K10. The knob acts as neo-centromere, moving to the poles earlier at both meiotic divisions, so that at second anaphase the knobbed daughter chromosomes pass preferentially to the outer products of the linear tetrad and since one of these, the basal megaspore, is the functional egg nucleus and hence there is an excess recovery of knobbed chromosomes. The phenomenon of meiotic drive has further been illustrated in case of monkey flower (Fig.16A) and silk eyed flies (Fig. 16B).

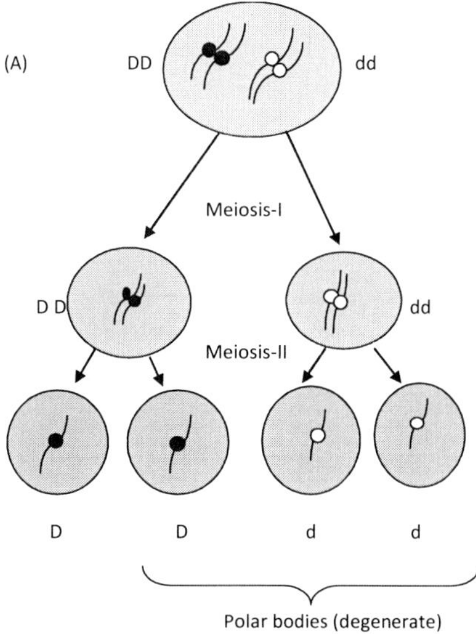

Fig. 16A: Mechanisms of meiotic drive. In figure 16A during female gametogenesis chromosome relegate rival chromosome to the polar bodies which are lost while drive chromosome enter the egg in monkey flower.

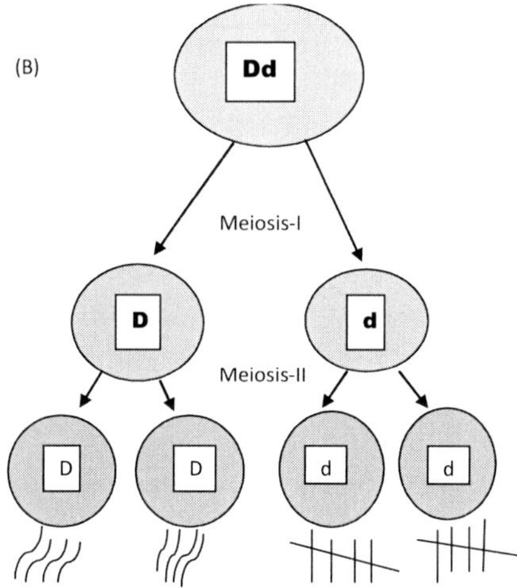

Fig. 16B: Male gametogenesis drive chromosome D causes death of sperms having rival chromosome (d) which lead to segregation distortion in silk eyed flies

B. Segregation of B chromosomes

In some plant species unpaired B chromosomes move at random to the poles in male meiosis but they move to micropylar pole at first anaphase in megaspore mother cell and are thus preferentially included in the functional egg nucleus leading to segregation distortation A marginal population of *Tettigidea lateralis* was found to be polymorphic with respect to a large mitotically stable B – chromosome. Male and female may carry one or two B-chromosomes In the males the frequency of individuals with one B and 2B was 33.8% and 2.9%, respectively. A small sample of females also had the similar frequencies In rye (*Secale cereale*) directed non- disjunction of the divided B chromosomes occurs during first pollen mitosis, as well as in the first mitosis during the development of the 8-nucleate embryosac, so that preferential transmission and increase of B -chromosome frequency takes place on both female and male sides (Muntzing, 1968). Since B- chromosomes are deleterious in large numbers the meiotic drive perhaps prevents their loss due to their positive effects during selection.

C. Intra plant variation

The orderly sequence of chromosome behavior during both of the cell divisions is designed to provide stability and genetical continuity between cell and organism generations. However, there are evidences where departure from this pattern occurs because of abnormal conditions tolerated by in many species of plants. The most common example of intraplant variation shown by cells in which, by successive replication cycles during interphase, the nuclei have two or more times as much DNA as the original nucleus. This endoreplication gives rise to polyteny in which chromatids remain laterally attached to give ribbon like chromosome with up to a thousand or more strands. These polyteny chromosomes are best known in the salivary glands of Diptera but they are also characteristic of endosperm, antipodal, haustorial and other cells in ovules of many plants.

The polyteny chromosomes in Diptera maintain a constant morphology throughout replication but in plants, atleast in the embryo suspensor cells of *Phaseolus*, they decondense and recondense during the replication cycle (Brady and Clutter, 1974). Endomitosis, in which the number of chromosomes increases, can also result from such replication. Endopolyploidy is particularly characteristic of non meristematic and non germline cells such as cortex, pith and xylem vessels, as well as in nutritive tissue like tapetum and endosperm. It occurs only in the cells which will not give rise new lineagesThe reasons for the regularity of this phenomenon are not clear, however, it may have some positive role in nutritive tissues.

A simple relationship exists between the chromosome number of somatic and germline cells. But in case of *Rosa candida* (2n=35) such simple relationship is violated as 14 chromosopmes out of 35 form 7 bivalents and 21 remain as univalents (7^{IIs} $+21^{Is}$) in megaspore mother cell.The 7 bivalents segregate normally while the 21 univalents collectively move to the micropylar pole and divide at second anaphase so that embryo develops from the nuclei with 28 chromosomes at the micropylar end of the linear tetrad. In pollen mother cells only the bivalents behave normally; and univalents divide irregularly and segregate at random so that about 90% of the pollen grains (gametes) have unbalanced chromosome number and only those with 7 univalents derived from 7II (bivalents) are functional. After fertilization, the original chromosome number (2n=35) is restored. Further more, cells having different chromosome number are too produced irregularly in plant tissues as a result of external factors such as wounding etc. violating the basic rule of meiotic division.

Synapsis and chiasmata

The pairing between homologous chromosomes during zygotene may begin simultaneously at several points along the chromosome or may be initiated at the centromere, near the telomere or at some combination of all these. Synapsis is valuable to determine the resemblance or homology between chromosomes brought together in hybrid combination/s. Although synapsis requires homology but in presence of certain genes in many plants even homologous chromosomes fail to pair (John and Lewis, 1965) and heterochromatic regions may show nonhomologous association. Synapsis is essential precursor for production of chiasmata formed between nonsister chromatids during pachytene. However, achiasmate meiosis is also known in plants, for example in *Trillium*, diploid *Lycopersicum x Solanum* hybrids and triploid *Allium amplectans*.The distribution of chiasmata does not occur at random, since the presence of one chiasma restricts the formation of another one in its proximity (positive interference) and the presence of heterochromatin seems to restrict chiasmata formation (Dyer, 1963) Furthermore, there are evidences that frequency and distribution of chiasmata are under genetic control. For example, males have fewer and more localized chiasmata than female meiotic cells in *Fritillaria, Lilium* and *Allium* (Fogweill, 1958 and Ved Brat, 1966and their total recombination frequency per bivalent in under strict genetic control. All critical data (John and Lewis, 1963 and 1965) support the original demonstration by Creighton and Mc Clintock (Fig.6) that chiasmata, involving non sister chromatids, are the first cytologically observed results of genetic recombination.

Molecular models of recombination and theories of crossing over

The term crossing over and recombination are generally used synonymously. However, **recombination** encompasses crossing over (both meiotic and mitotic), gene conversion, exchange between sister chromatids and repair of damaged DNA. Homologous recombination models fall into three classes:

1. Copy choice model

In copy choice model recombination occurs during the synthesis of daughter DNA molecules DNA replication starts with one parental DNA template and then switches to the second parental molecule, resulting in the synthesis of recombinant daughter DNAs containing sequences homologous to both parents. (Fig.17). According to this theory only two strands involve in crossing over, however, as a known fact crossing over may be three strand or four strand involving different strands (Fig.12 A, B, C, D).

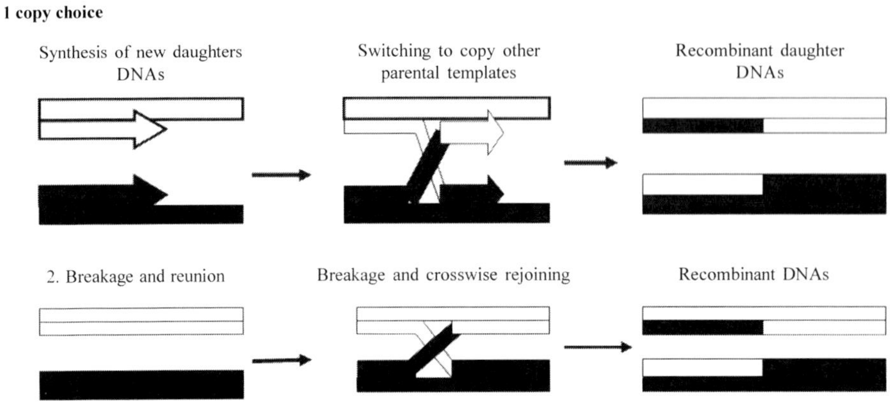

Fig. 17: Recombination models. 1. Copy choice recombination occurs during the synthesis of daughter DNA molecules. DNA replication starts with one partental DNA template and then switches to second parental molecule, producing recombinant daughter DNAs containing sequences homologous to both parents. 2. In breakage and reunion recombination occurs as result of breakage and crosswise reunion of parental DNA molecules

2. Breakage and reunion

In breakage and reunion model recombination occurs in the absence of DNA replication: DNA strands are broken, exchanged between duplexes & relegated (Fig.17-2). The theory of breakage and reunion was proposed by Darlington in 1935 and according to this theory recombination occurs as a result of breakage (chromatids) and there after crosswise rejoining of parenatal molecules. If breakage occurs in more than one chromatid and they reunite with other chromatids, it would lead to recombination. However, if it reunites with a sister chromatid, it would have no genetic consequence. The copy choice model proposes that the recombinant molecule is generated during DNA synthesis, as a result of copying first one parental DNA and then switching to copy a different template but breakage and reunion model suggested recombination to occur from the breakage and reunion of two parental DNA molecules rather than by synthesis of new DNA. The breakage and reunion theory was resolved in 1961 by Matthew Meselson and Jean Weigle by simultaneously infecting E.coli with two strains of λ carrying different genetic markers (A and B).To determine whether recombination involved breakage and rejoining of parental DNAs, one of the parental viruses was grown in medium containing the heavy isotopes of carbon (^{13}C) and nitrogen (^{15}N), while the other was grown in medium containing the normal light isotopes (^{12}C & ^{14}N) which resulted into parental viruses to have different densities which could be separated by equilibrium density centrifugation in a CsCl gradient. *E.coli* were then simultaneously infected \with these differentially labeled (heavy and light) parental visuses under conditions where replication was inhibited; and the progeny viruses thus produced were analyzed for both their density and genetic markersThe important experimental finding was that co-infection produced genetic recombinant viruses having intermediate densities rather than heavy or light, indicating that they had acquired the DNA from both parents, due to exchange event between two viral DNAs having different markers. (Fig 18) as proposed by breakage and rejoining theory.

3. Hybrid models

These models combine features of both breakage and reunion model and copy choice model. However, all these three types of recombination occur under different circumstances, ie, homologous recombination during meiosis primarily occurs by strand breakage and reunion, where as DNA repair by recombination involves copy choice or hybrid mechanisms.

Types of intramolecular recombination

Creighton and Mc clintock (1931) cytologically established that physical exchanges between chromosomes are accompanied by recombination of markers (genes) located on the homologous chromosomes and how do such chromosomal exchanges occur at molecular level was established by a seminal model of Robin Holliday(1964). An important consequence of the Holliday Model was that it predicted a structure which, along with its modifications, is the basis of present day thinking about molecular recombination. The model provides a molecular basis for the association between aberrant segregation and crossing over. Gene-conversion and post meiotic segregation are assumed to result from repair of mismatches, or lack there of, on one or both chromosomes in a symmetric structure containing heteroduplex DNA.

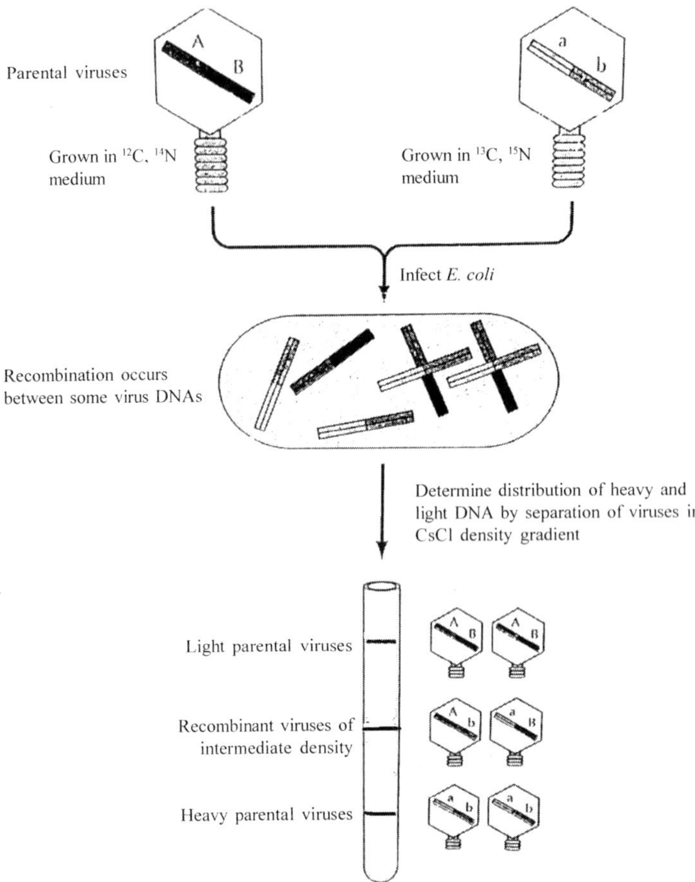

Fig. 18: Experimental evidence of breakage and reunion model of recombination. Genetically distinct ë viruses were grown in medium containing either light isotops of carbon and nitrogen(^{12}C ^{14}N) or heavy (^{13}C ^{15}N) to density label their DNAs. E. coli were simultaneously infected under condition in which replication was inhibited. Thus produced progeny viruses were harvested and analysed by equilibrium centrifugation in a CsCl gradient to analyse the density of genetic recombinants. The recombinants having intermediate densities and exchanged markers (A, b and aB) were observed, proving that they had acquired DNA from both parents by breakage and reunion mechanism (Meselson and Weigle, 1961)

1. Site specific recombination

It is not dependent upon homology between recombining partners as in case of homologous recombination. The proteins mediating this process recognize short specific DNA sequences in the donor and recipient molecules, and interaction between these proteins facilitates recombination. Homology often exits between the donor and recipient sites because the same recombinase protein binds the both recognition sites. The basic difference between homologous recombination and site specific recombination is that the homologous recombination occurs at any extensive region of sequence homology while site specific recombination occurs between specific DNA sequences which are usually a short stretch of DNA. The example of site specific recombination is integration and excision of bacteriophage λ when it infects *E. coli*. It may either replicate inside ***E.coli*** and cause lysis or may integrate into bacterial chromosome forming a prophage to be maintained as part of *E.coli genome*. However, integration of bacteriophage λ can also be reversed under appropriate conditions resulting its excision and there after initian of lytic viral replication.The DNA of *E.coli* and bacteriophage λ recombine at specific sites called **attachment sites (att)**. The integration of λ DNA involves recombination between att sites of the phage **(Att P)** and *E.coli* (att B) which are about 240 and 25 nucleotides long, respectively. The process of integration is mediated by a protein of bacteriophage λ called integrase **(Int)** by binding to both **Att P** and **Att B**. sequences.Integrase (Int) first binds to att P, forming a complex in which **att P** DNA is wrapped around multiple copies of the Int protein. The initial **att P** complex binds to the **att B** aligning the phage and bacterial **att** site. The phage and bacterium then exchange the strands within a 15 –nucleotide core sequence shared by attB and att P (Fig. 19 & 20). The Int protein causes staggered cuts within the core homology region of attB and attP, catalyzes strand exchange and the seals the broken ends by lygases and thus integrating the lambda DNA into ***E.coli*** chromosome (Fig. 19 & 20). The excission or reversal of viral DNA also caused by Int protein.

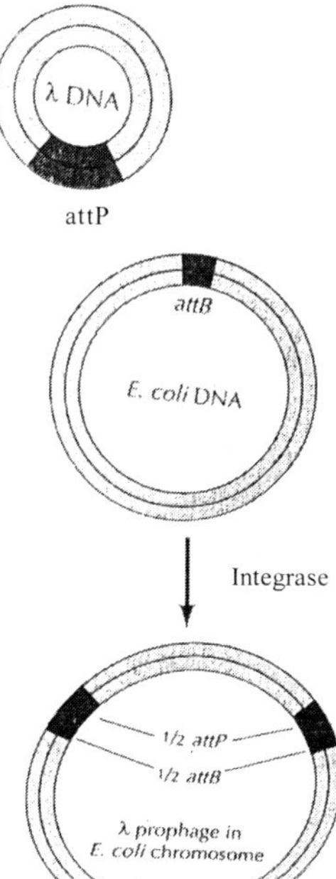

Fig. 19: Site specific recombination between *E.coli* and msul lamdab bacteriophage. Integration of is given in upper fig. DNA is caused by recombination between specific sequences in the l and *E.coli* genomes, ie, attP and att B, respectively. Virus encoded enzyme integrase (Int) which catalyses the process by recognizing the att P and att B sequences.

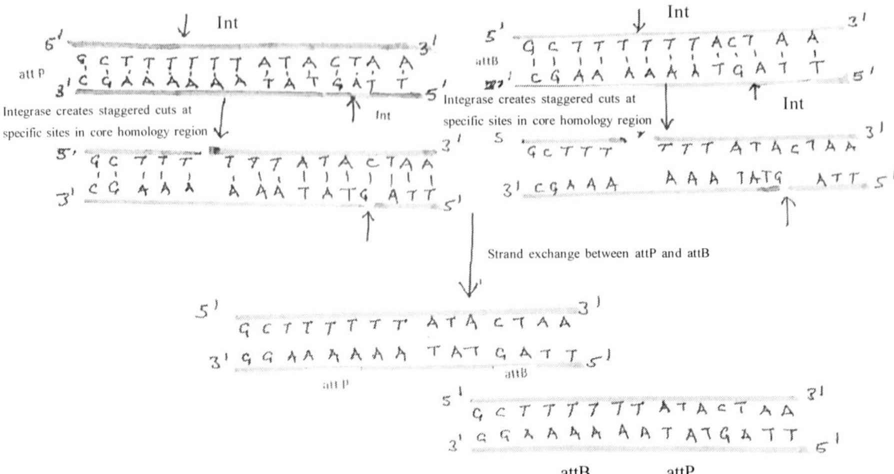

Fig. 20: Occurrence of site specific recombination between *E.coli* and bacteriophage λ involving common homologous core sequence of att P and attB. Integrase (Int) cleaves at specific sites within core sequence to create staggered single stranded DNA tails. It also catalyses strand exchange and ligation, leading to recombination between att P and attB and λ integration into *E.coli*

2. Illegitimate or unequal recombination

It requires little or no homology between combining partners and results from aberrant cellular processes like illegitimnate end joining, strand slipping and looping during replication. The bar and ultra bar eye in drosophila are produced by the phenomenon of unequal crossing over.

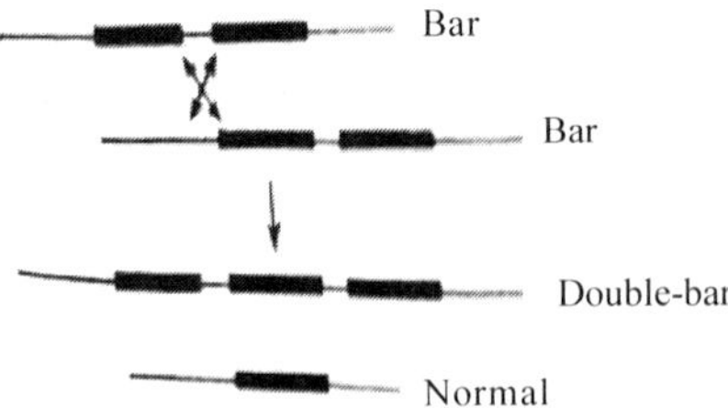

Fig. 21: Illegitimate or unequal recombination between duplicated regions causing double bar eye in Drosophila.

3. Transposition

It does not require homology between recombing partners. The proteins mediating the process (transposases, integrases) recognize short specificDNA sequences in one of the recombining partners only, which is a transposable element (the site of recognition is usually at the junction between the transposable element and the host DNA. The recipient site is usually relatively non specific in sequence where recombination integrates the transposable element into the host DNA.

4. Artificial recombination

Artificial recombination is carried out by DNA ligation using purified enzymes and substrate in vitro. The notable example of such a recombination is synthesis of recombinant DNA molecule.

5. Homologous recombination

It requires homology between combining partners.The proteins mediating this process (e.g.RecA in *E. coli*) are not sequence specific but are homology dependent.Long regions of homology are involved during meiotic division.However, the findings that recombination occurs by breakage and reunion raises a critical question that how two parental DNA molecules are broken precisely at the same point and rejoin without any deficiency or duplication (mutation) of nucleotids at the break point.To avoid any addition or deletion of nucleotides during homologous recombination the alignment of two DNA molecules is provided by base pairing between complementary DNA strands (Fig 22). Overlappinjg single strands are exchanged between homologous DNA molecules leading to formation of a heteroduplex region, in which the two strands of the recombinant double helix are derived from different parents. If the heteroduplex region contains a genetic difference resulting a single progeny DNA molecule that contains two genetic markers. In some such cases mispaired bases in a heteroduplex may be identified and corrected by mismatch repair system. The genetic evidence of such identification and correction in heteroduplex was obtained in studies of recombination in fungi and bacteria which led to the development of a molecular model to explain complex recombination. This molecular model is known as "HollidayModel" which along with some modifications is providing basis for recombination mechanisms. The original model presumes that to initiate the recombination two DNA molecules are nicked at same position. The nicked DNA strands are partially unwound, and each strand invades the other molecule by pairing with the complementary unbroken strand. Ligation of the broken strands leads to formation of crossed–strand intermediate, termed as **Holliday Junction.** Thus formed Holliday Junction can be resolved in several ways to yield distinct genetic consequences leading to complex recombinations.The first modification of this model was proposed in 1975 which suggested that nicking occurs only in one parental molecule instead of two. The nicked strand is displaced, and the resulting single strand invades the other parental molecule by homologous base pairing. This process produces a displaced loop of DNA, which can then be cleaved and joined to the other parental molecule leading to formation of Holliday Junction which can be again resolved into recombinant and non-recombinant heteroduplex. The third modification suggested that recombination is initiated by double strand nick rather than a single strand nick and exonuleases at the site of break then generate single- stranded tails, which can invade a homologous double-stranded molecule and thus again forming a Holliday Junction. This double strand nick repair model is reported to be applicable to meiotic recombination in yeast.

The basic feature of all these models account for the initial stages of recombination between two DNA molecules and establishment of **crossed strand Holliday Junction** for recombination in eukaryotes.The details of the "Holliday Model" are given below

Creighton and Mc clintock (1931) cytologically established that physical exchanges between chromosomes are accompanied by recombination of markers (genes) located on the homologous chromosomes. But how do such chromosomal exchanges occur at molecular level was established by a seminal model of Robin Holliday(1964). An important consequence of the Holliday Model was that it predicted a structure which, along with its modifications is the basis of contemporary thinking about molecular recombination.The model provides a molecular basis for the association between aberrant segregation and crossing over.Gene conversion and post meiotic segregation are assumed to result from repair of mismatches, or lack there of, on one or both chromosomes in a symmetric structure containing heteroduplex DNA.

Salient features of Holliday Model for homologous recombination

The homologous recombination is divided into four stages: synapsis, strand transfer repair and resolution (Fig. 22).

1. **Synapsis**, homologous duplexes are aligned due to homology. An endonuclease creates single strand nicks into each of the strands.
2. **Strand transfer**: A single DNA strand is transferred (strand invasion) from one duplex to the other. The first strand transfer marks the initiation of recombination as it invades the homologous duplex and (if the recipient duplex is intact) displaces the resident strand. This process may generate a short region of heteroduplex DNA. Since duplex DNA comprised of strands from different parental molecules which may contain reflecting sequence differences (different alleles) in the parental duplexes. If the recipient duplex is intact, the displaced resident strand is able to pair with the free strand of initiating duplex. The two transferred strands cross each other, forming a structure termed as **Holliday junction**, **cross bridge** or **Holliday intermediate** (Fig.22). The site of Holliday junction may move in relation to its original position by progressive strand exchange between duplexes This process is termed as **branch migration** which may increase or decrease the amount of heteroduplex DNA.

The Holliday model of homologous recombination between intact duplexes

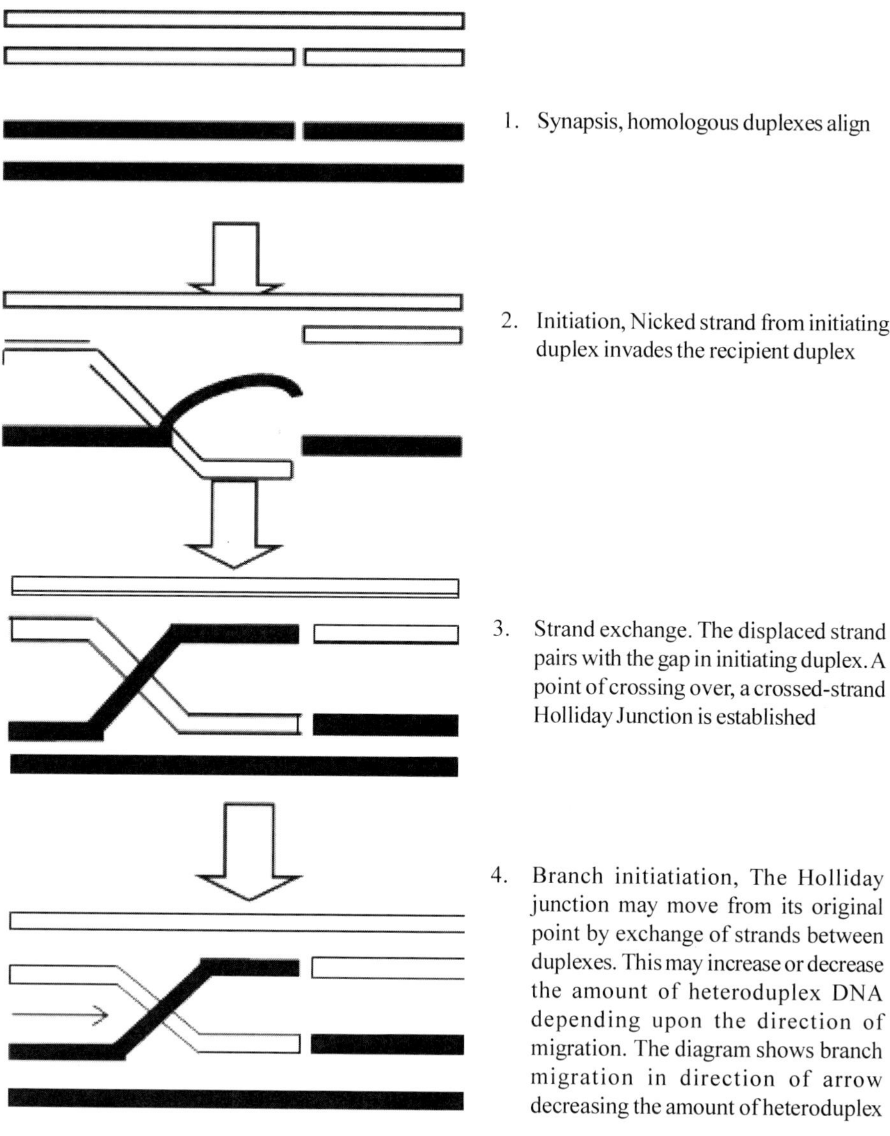

Fig: 22: The Holliday model ofhomologous recombination between intact duplexes, showing synapsis and strand transfer stages. The classic model involves duplexes with nicks in the equivalent positions. However, this is an unusual situation —normally, only one of the duplexes is nicked and the invading strand displaces the resident strand.

3. **Repair and resolution**: The cutting process which separates the two participating DNA molecules is called **resolution.** These processes do not occur in fixed order as it depends upon the recombining partners and the availability of appropriate enzymes. Repair refers to three different processes.Firstly the recombining duplexes remain intact (i.e. there is no genetic information missing from either duplex) and repair simply involves relegation of the broken strands. This is a **conservative type ofrecombination**. However, intermidiate Holliday structure may also beresolved in either of the two planes to generate one of two products – a **patch** of heteroduplex DNA in a non- recombinant background or a **splice**of heteroduplex DNA with recombination of flanking markers A and B(Fig. 23).

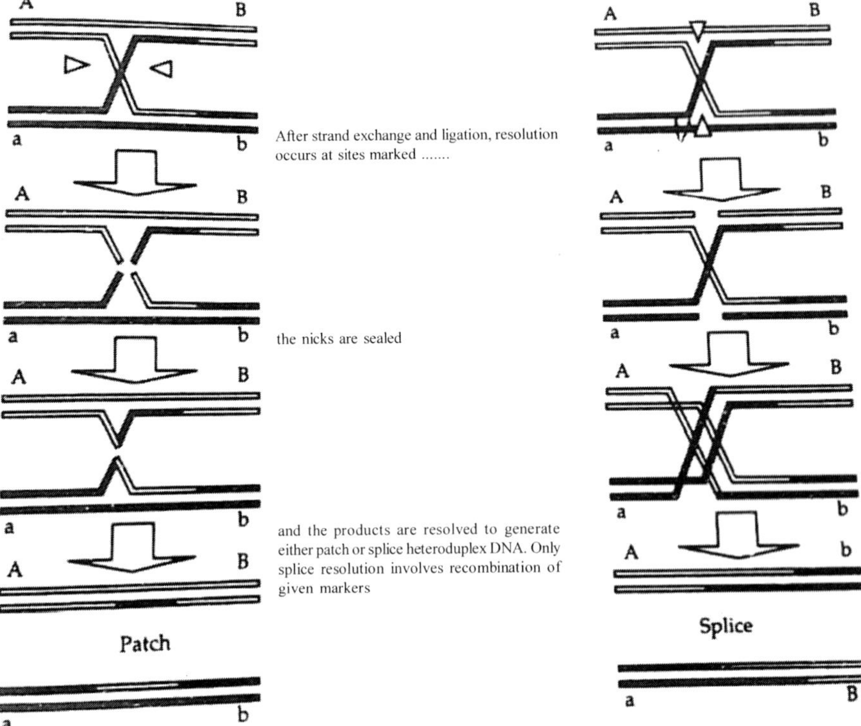

Fig. 23: Resolution (separation of the two participating DNA molecules) of the Holliday junction in either of two planes, generating different products. Only one resolution pathway generates a recombinant molecule for markers A and B, although both pathways generate a region of heteroduplex DNA, known as a splatch (patch and spice).

However, if genetic information is missing from either duplex (i.e. if there is a single strand gap, or a break) DNA repair synthesis replaces the missing information using information from homologous duplex as a template, where as recombination including the synthesis of new DNA is nonconservative.In the extreme cases where an entire chromosome segment is missing, resolution of Holliday junction yields a replication fork which can duplicate the missing segment. The third type of DNA repair, mismatch repair of heteroduplex DNA is generally random in its direction (post replicative mismatch repair) and may cause gene conversions. Reciprocal recombination is both symmetrical and conservative, i.e., information is exchanged between duplexes and nothing is lost or gained. Thus, if recombination occurs between parental chromosomes carrying alleles AB and ab, the reciprocal recombinants are **Ab** and **aB.** However, non reciprocal recombination involves unidirectional transfer of information. An example of nonreciprocal recombination resulting from the above cross would be **Ab** and **AB.** Where information has been transferred unidirectionally from A to the a allele where a allele has been lost. nThis type of recombination lead to **gene conversion** because one allele has been converted to other one. Gene conversion or non reciprocal recombination occurs under two circumstances.

Firstly consider the fate of the heteroduplex DNA which arises during strand invasion and branch migration. When Holliday junction is resolved, heteroduplex DNA, which contains base mismatches, can be subjected to mismatch repair. The repair can be in either direction because meiotic strand exchange occurs well after replication and the cell can no longer discriminate between thc parent and daughter strand. However, for gene conversion event, a locus must be close to the site of strand exchange within the scope of branch migration, and the phenomenon, therefore, is most commonly observed when recombination of closely linked marker considerd. Because repair is non directional, gene conversion may be obscured in a population of meiotic products where equal representatives of each repair process may be found. However, it is easily observed in fungal tetrads, which represent individual meioses (tetrad analysis).

Secondly gene coversion may occur during the recombinational repair of gaps and breaks. Single stranded gaps may be filled by a strand from a homologous duplex prior to repair which generates a heteroduplex DNA that may be repaired in either direction. Double stranded breaks are often targets for exonucleases, causing substantial loss of information prior to repair In this case conversion is always in the same direction, i.e., towards the homologous undamaged chromosome because this is the only information available.

Techniques of Karyotyping, chromosome banding and painting

Any individual plant or an animal is characterized by a set of chromosomes which have certain constant features. These features include chromosome number, size and shape of individual chromosome/s. The display of full chromosome complement of a particular species comprising of these features is termed as

karyotype which may be represented by a diagram called **ideogram**, where chromosomes of haploid set of an organism are arranged in series according to decreasing order of size.For karyotypic analysis, the number and morphology of chromosomes from well spread out mitotic preparations is utilized.The measurement of chromosomes is made with the help of ocular micrometer. The chromosome length, positions of primary and secondary constrictions and satellites are determined based on somatic metaphase chromosomes especially from root tips of plants (Fig.25). Furthermore, differences in total length, arm ratio, size and shape of knob, satellite and nucleolus organizer and prominent chromomeres in certain regions also serve to identify and distinguish many of the chromosomes.Usually diploid karyotype is prepared by cutting out pairs of homologous chromosomes from a metaphase stage of root tip mitosis photograph which are arranged in decreasing order of size (Fig.24) where as in case of haploid nucleus only one representative of each chromosome type is presented. The examination of chromosome pairs reveal that each of these pairs is of different size and the position of centromere also differs in these pairs. Some are metacentric having medially positioned centromere having about equal length of both arms; others are acrocentric where one arm is longer than other and rest are intermediate of these two categories (Fig.25-2). Karyotypes are useful in identification of chromosomal aberrations, chromosome number or morphology.They may also be utilized in establishing the evolutionary relationships between different species. Furthermore, the allotment of genetic material to more than one major chromosome is a feature that distinguishes eukaryotes from prokaryotes. The multi chromosome state appears to facilitate the packaging and distribution of enormous amount of DNA in higher organisms.The meiosis and sexual fusion in multi chromosome organisms also create new combinations of genes which ultimately lead to diversification and genotypic flexibility for evolution.

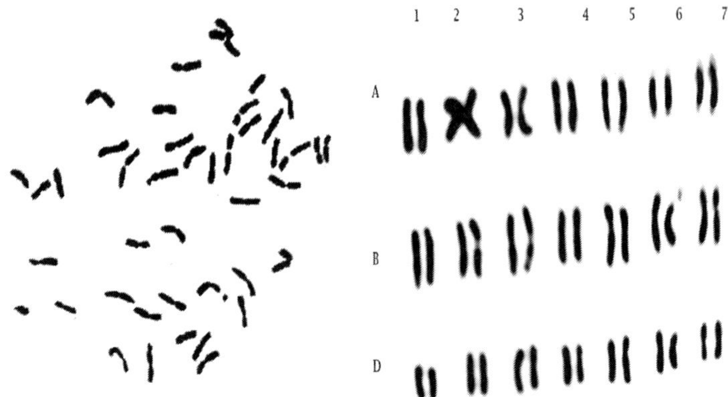

Fig. 24: Metaphase chromosome spread of wheat 2n=42 and karyotype.

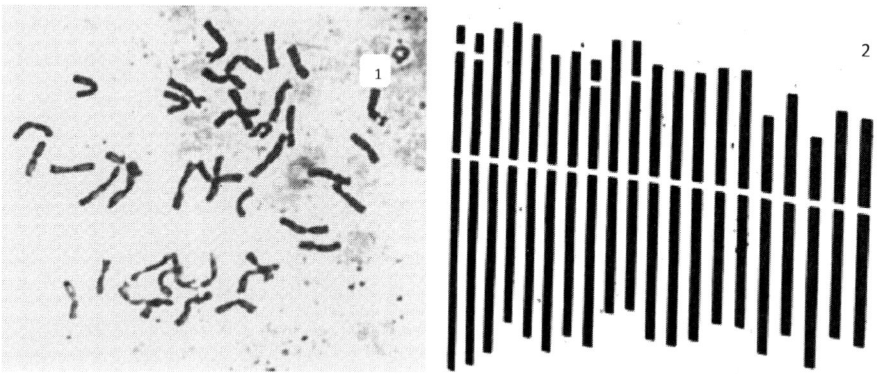

Fig. 25: Metaphase spread (1) and Ideogram (Fig.2) and Chromosome lengths and arm ratios of hexaploid triticale Bronco-90 (2n=42) (Table 3).

Chromosome	Chromosome length (μ)	Arm Ratio	Satellite length (μ)
SAT1	10.03	2.04	0.60
SAT2	9.36	2.10	0.58
SAT3	7.95	2.21	0.62
SA41	7.45	1.47	1.00
M_1	8.61	1.10	
M_2	7.52	1.02	
M_3	6.85	1.15	
M_4	6.09	1.14	
M_5	5.61	1.14	
SM_1	9.31	1.44	
SM_2	8.57	1.27	
SM_3	8.32	1.64	
SM_4	7.98	1.41	
SM_5	7.45	1.39	
SM_6	7.40	1.47	
SM_7	7.24	1.38	
SM_8	6.21	1.30	
SM_9	5.76	1.33	
SM_{10}	5.69	1.50	
ST 1	6.22	2.07	
ST2	5.81	2.52	

The metaphase chromosomes were identified and categorized on the basis of size and morphology in four groups.

1. Satellited chromosomes. Four pairs of satellite chromosomes were observed in full complement of hexaploid triticale. One chromosome designated SAT 4 has the longest satellite.
2. Median chromosomes: All chromosomes having arm ratio ranging from 1 to 1.25 were grouped in this category. The length of these chromosomes ranged from 8.61to 5.61 µ. There were five pairs of such median chromosomes.
3. Sub-Median chromosomes: The chromosomes having arm ratio from 1.25 to 1.75 µ were grouped in this category. The chromosome length ranged from 9.31 to 5.69 µ. There were 10 pair of such sub-median chromosomes in whole complement.
4. Sub- terminal chromosomes: The chromosome having arm ratio above 2.00 were classified as sub-terminal chromosomes. There were 2 pairs of such chromosomes.

Feulgen staining for nomal karyotype (Mitotic chromosome)

Young root tips are prefixed for 2-3 hours at 10°C to 14°C in a saturated solution of 8-hydroxyquinoline or para-dichlorobenzene and after washing in water they are fixed in acetic acid: ethyle alcohol (1:3) for about 24 hours. The cells of root tips are subjected to washing with 1N HCl and to mild hydrolysis in 1 N HCl at 60°C for 10-12 minutes in water bath or oven. After hydrolysis, root tips are washed in water to remove the traces of acid and kept in Feulgen stain for half an hour in dark to take them bright magneta color. The stained root tips portions are taken to a clear slide for squashing and chromosomal observations after adding 1or 2 drops of aceto- carmine. The treatment of mild hydrolysis produces a free aldehyde group in deoxyribose molecules. The aldehyde then reacts with a chemical known as Schiff's reagent (basic fuchsin bleached with sulfurous acid) to give magneta color.

Chromosome banding

If the karyotypes are constructed by conventially stained preparations, it may be virtually impossible to distinguish many of the sister chromatid pairs from one another. To overcome such a problem, some other staining procedures have been developed and applied to reveal discrete bands along the lengths of sister chromatid of a chromosome. These bands readily permit the identification of individual chromatid pairs and the matching of homologues with similar banding patterns.

The chromosome banding technique, as it stands in present times, has revolutionized cytogenetics. It has made possible precise identification of individual chromosomes or and even parts of chromosomes, their structural and molecular organization, chromosome change/s during evolution, studies in chromosaome polymorphim, detection of a structural karyotypic changes, genomic analysis in allopolyploids, aneuploid identification and detection of inter-specific translocation. The chromosome bands either can be observed along the full length of chromosome as G-, Q-, and R- bands or restricted to a certain area of chromosome with specific bands as C-bands.

Many staining protocols (listed below) have been developed in various laboratories to produce banded chromosomes:

1. **C-banding**: A giemsa based technique which includes incubation with barium hydroxide to produce c-bands. The C bands correspond to non coding constitutive heterochromatin surrounding the centromeres and hence useful to determine the position of centromere. The technique was developed by M.L.Pardue and J.Gall following *in situ* hybridization of mouse satellite DNA with complementary RNA. The detection of centromeric heterochromatin is based upon the denser stainability of giemsa stain. In C-band technique, air dried chromosome preparations are treated with alkali (i.e. saturated solution of $Ba(OH)_2$ for 5 minutes (denaturation) followed by renaturation, in 2SSC (0.3 sodium chloride+0.03M sodium citrate) for 1-2 hours at 60°C and subsequently stained with diluted solution of giemsa in phosphate buffer. The technique produces dark bands (c-bands) at the centromeric regions of chromosomes (constitutive heterochromatin). This method, beside centromeric bands, also gives bands on the chromosome arms.

2. **Q-banding**: A banding technique using the fluorescent dye quinacrine mustard staining and observations with ultraviolet light. This produces a similar pattern to G-banding, but also causes areas of hetereochromatin to fluoresce brightly alternating with dull fluorescence. The Q banding is very useful to identify Y-chromosome since it becomes brightly fluorescent, both in interphase nucleus and in metaphase.

3. **G-banding**: The G bands (from Giemsa) have the same locaton as the Q bands and do not require fluorescent microscopy. Many techniques are available, each involving some pretreatment of the chromosomes. In the acid-saline-Giemsa (ASG) technique cells are incubated in citric acid and NaCl for 1 hour at 60°C and are then treated with the Giemsa stain. The technique is mainly confined to animal system.

4. **R- banding**: A Giemsa based technique including a heat- denaturation step which results in a reversed banding pattern from that obtained with conventional G- banding. The air dried preparations of chromosomes (animal) after controlled incubation at 87°C in phosphate buffer for 10-12 minutes, produces paler staining regions which are easily visible under phase contrast

microscope This technique is particularly useful to identify the terminal deletions of chromosomes.

5. **Replication banding**: The technique involves the incorporation of bromodeoxyuridine into replicating DNA. A brief pulse during the S phase allows replication timing to be investigated and if prolonged, generates a banding pattern similar to Giemsa stain, suggesting G-bands correlated to replication time zones. Incorporation over two rounds of replication allows sister chromatids to be discriminated on the basis that one has bromodeoxyuridine incorporated in both strands and other in only one strand. Chromatids then stain differently in the presence of Giemsa stain and the fluorescent dye (Hoechst 33258). This technique is known as harlequin staining and is useful for the detection of chromatid exchange

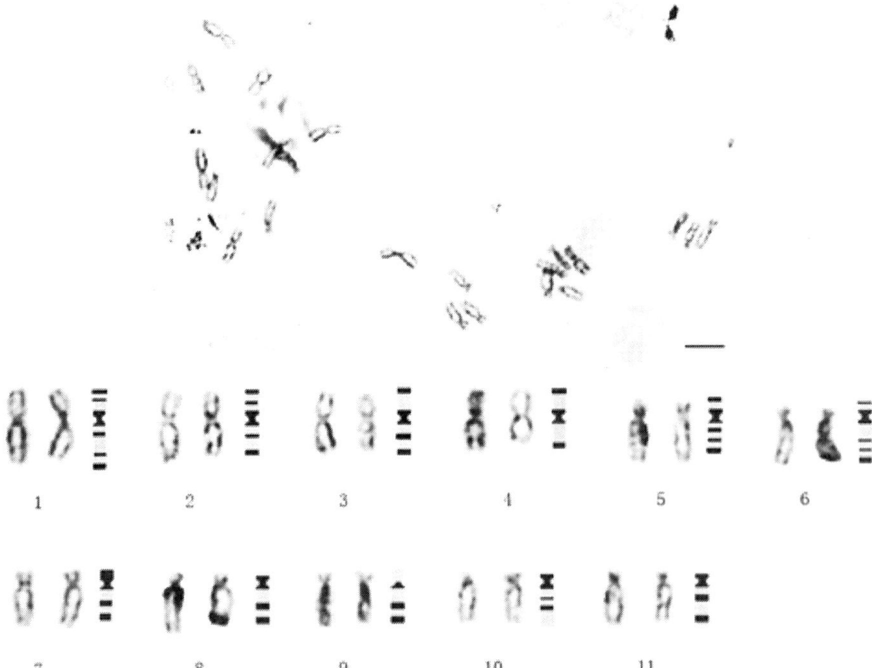

Fig. 26 : C-banding karyotype of *E. purpurea* (W. Jiang et al. 2016; Hereditas 2016;153:14)

(A) *In situ* hybridization (ISH)

This technique is used to precisely localize the position of a known DNA sequence on a chromosome in a interphase or metaphase mitotic spread.The basic principle of the technique is hydrogen bonding between nucleotides in double helix of DNA(A=T) (C≡G) which provides a remarkable stability to double helix. However these bonds can be broken with heat or chemicals (denaturation) to unwind the helix. Under favorable condition, the helix is able to reform. The ability of DNA to reform and renature is exploited in molecular hybridization where labeled DNA or RNA sequence is used as a probe to identify/quantify the naturally occurring counterpart of the sequence in situ (i.e.) in their natural positions within a chromosome). The brief technique of ISH is as follows:

1. The DNA within the cell is denatured (converted into single stranded one) by treating the cells in mitotic spread.
2. The squashed cells are then incubated in a solution of labeled DNA whose position on the chromosome is to be ascertained. The following types of probes are used for this purpose:
   - Double-stranded DNA (dsDNA) probes
   - Single-stranded DNA (ssDNA) probes
   - RNA probes (riboprobes)
   - Synthetic oligonucleotides (PNA, LNA)

   The probes are labeled by radioactive and non- radioactive labels.
   - **Radioactive isotopes**
   - ^{32}P
   - ^{35}S
   - ^{3}H
   - **Non-radioactive labels**
   - biotin
   - digoxigenin
3. The slides are pretreated and chromosomal DNA and probes are denatured according to standard procedure and their mix is prepared.
4. The labeled probe is placed on the slide having chromosomal spread, put up the cover slip with rubber cement and incubated for 6-14 hours at prescribed temperature for various probes.
5. Remove the cover slip and wash off the hybridization mix in 2xSSC (Sodium chloride saline citrate as prescribed.
6. In case of radio actively labeled probe, the slides are dehydrated with ethyl alcohol (70 % or 95 %), air dried and subjected for autoradiography to locate the position of known DNA sequence on a specific chromosome. However, standard procedure may vary slightly according to use of various probes.

(B) Fluorescence *in situ* Hybrididation (FISH)

Joseph Gall and Mary lou Pardue (1980) realized the potential of molecular hybridization to identify the exact location of DNA sequences within a chromosome/s. Ever since those original observations, many refinements have been made to enhance the sensitivity and versatility of the procedure to the extant that in present times it is an essential tool in cytogenetics.Traditionally in (as above) *in situ* hybridization radio active probes are used but in FISH technique radio active probes are replaced by fluorescent ones to detect DNA sequences and the process is commonly referred as to FISH (fluorescence *in situ* hybridization).Apart from its speed and efficiency FISH has the advantage that probes with fluorochromes can be used to identify different targets at the same time with different colors, allowing gene order to be determined. FISH to metaphase chromosomes gives a resolution of 1-10 Mbp.

The first step in the process is to make either a fluorescent copy of the probe sequence or a modified copy of the probe sequence that can be rendered fluorescent later in procedure. The probes are labeled with biotin, digoxigenin, dinitrophenyl, DNP, aminoacetyl fluorine or mercury etc. Next, before any hybridization occur, both target (nucleus) and probe sequences must be denatured (melting) with heat or chemicals. The denaturation step is essential in order to form new hydrogen bonds between the target and the probe during their subsequent hybridization, where probe and target sequence are mixed together and probe specifically hybridize to its complementary sequence on the chromosome.Since probe is already fluorescent, the hybrids thus formed between probes and chromosomal targets, can be visualized and detected using a fluorescent microscope.

Application of FISH

1. To identify gene location

FISH is a powerful tool to locate position of cloned genes on metaphase chromosome by hybridizing normal metaphase chromosome with cloned DNA sequence by their characteristic banding patterns at hybridization sites on homologous chromosomes. The FISH has also been used to detect repeated sequences in wheat (Mukai *et al.*, 1993), barley (Leitch, 1993), rye (Albini, 1992), tobacco (Kenton *et al.*, 1993) and tomato(Zhong *et al.*, 1996).

2. Detection of chromosomal abnormalities using karyotypes and FISH

The chromosomal abnormalties, i.e., duplications, deletions and translocations can be detected by using FISH vis-a-vis standard kayotypes.

3. Chromosome painting

The cytogeneticists now have the option of using multifluor FISH or spectral kayotyping to quickly scan a set of metaphase chromosomes for potential rearrangements (Speicher et al. 1996; Schrock et al. 1996). Multifluor FISH generates a karyotype in which each chromosome appears to be painted with a different color. The term chromosome painting was coined by Pinkel et al. in 1988 to describe the use of complex mixtures of DNA sequences, representative of a single chromosome using in situ hybridization in metaphase spreads and interphase nuclei. The basic elements of chromosome painting are are shown in the figure (Fig.27):

1. The basic material for the production of chromosome paint is highly purified, flow sorted chromosomes.
2. Short sequences are generated from the sorted material (chromosome) and amplified by enzymatic digestion and molecular cloning into a bacterial vector or use of the polymerase chain reaction (PCR). The extracted cloned DNA or the PCR products are modified with biotin labeled nucleotides by nick translation and hybridized on to the metaphase spreads and interphase nuclei, duplexes form between complimentary sequences in the chromosome paint and target material along the length of chromosome type from which the paint was originally prepared.
3. Detection of hybridized paint with fluoresceinisothiocynate(FITC) conjugated with avidin results in direct visualization of clearly bright fluorescent chromosome.

The key to the successful specific hybridization of these libraries was the use of chromosomal in situ suppression (CISS) hybridization techniques.DNA cloned from any flow sorted chromosome will contain sequences shared with other chromosomes. Labelled and used directly as a probe on to metaphase spreads, such sequences would hybridize not only to the original sorted chromosome type but also to complementary regions of other chromosomes. To avoid such undesirable hybridization, the labeled library is mixed with an excess of unlabelled genomic or Cot 1 competitor DNA, denatured and then allowed to re-anneal for a short time, the first duplexes to form are predominantly from repetitive sequences shared between probe and the competitor. The single stranded probe remaining after this period is predominantly from sequences unique to the sorted chromosome and when applied to the metaphase spreads specific chromosomal hybridization is visualized. The second important point that determines the quality of chromosome paint is the purity of sorted chromosomes.

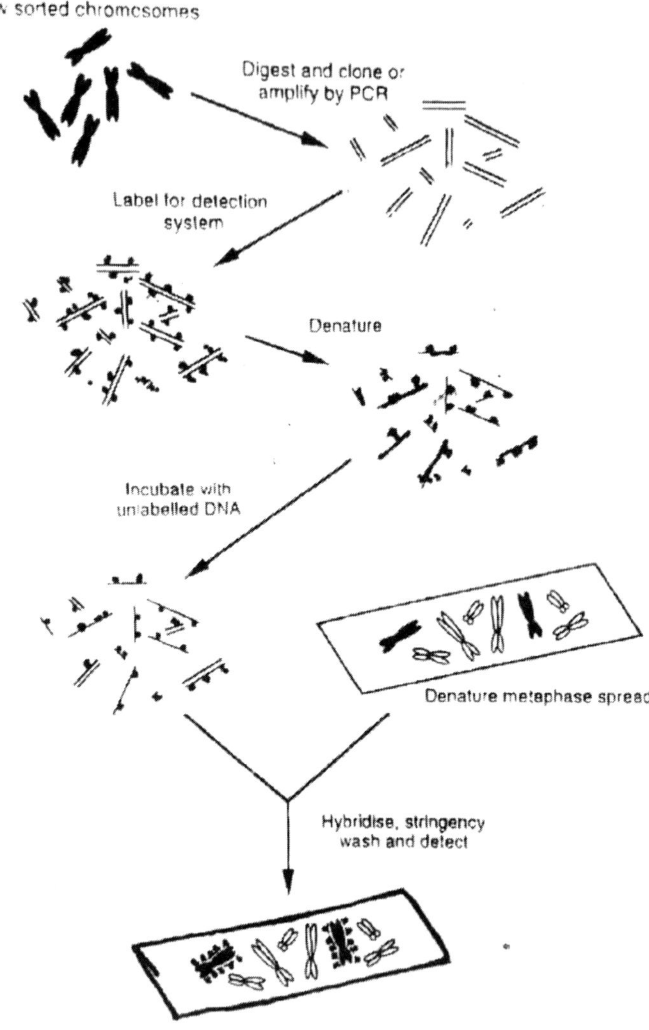

Fig. 27: Generalized procedure of chromosome painting

References

1. Albini, S.M. and Schwarzacher, T., (1992). Genome 35: 551-559.
2. Brady, T. and Clutter, M.E. (1974). Chromosoma (Berl.) 45: 63-79.
3. Creighton, H.B. and Mc Clintock, B. (1931). Proc. Nat. Acad. Sci. U.S.A. 17: 492-497.
4. Dyer, A.F. (1963). Chromosoma (Berl.) 13: 545-576.
5. Evans, G.M. and Rees, H. (1966). Exp. Cell Res. 44: 150-160.
6. Fogwill, M. (1958). Chromosoma (Berl.) 9: 493-504.

7. Hotta, Y; Parchman, L.G. and Stern, H. (1968). Pro. Nat. Acad. Sci. U.S.A. 60: 575-582.
8. John, B.and Lewis, K.R. (1965). Protoplasmatologia VI, F11-335.
9. John, B. and Lewis, K.R. (1968). Protoplasmatologia VI A:1-206. Springer –verlag, Wien.
10. John, B. and Lewis, K.R. (1969). Protoplasmatologia VI B: 1-125.
11. Kenton, A.; Parokonny, Y.Y. Gleba and Bennet, M.D. (1993). Mol. Gen. Genet. 240: 159.
12. La Cour, L.F. and Wells, B. (1974) Caryologia 27: 83-92.
13. Leitch, J. and Heslop-Harrison, J.S. (1993). Genome 36(3): 517-523.
14. Lewis, K.R. and John, B.(1963). Chromosome Marker, Churchill, London.
15. Maziz, D. (1961). The Cell. 3: 77-412. Academic Press, New York.
16. Morgan, T.H. (1910). Science 32: 120-122.
17. Mukai, Y., Friebe, B., Hatchett, M.and Yamamoto, M. (1993). Chromosoma 102(2): 88-95.
18. Muntzing, A. (1968). Hereditas (Lund) 59: 298-302.
19. Nagl, W. and Ehrendorfer, F.(1974). Plant Syst. Evol. 123: 35-54.
20. Schrock, E., du Manoir, S. Veldman, T., Schoell, B.Wienberg, J. Ferguson-Smith, M.A. Ning, Y. Ledbetter, D.H., Bar-Am, I.Soenksen, D. Garini, Y.and Ried, T. (1996). Science Jul. 26; 273(5274): 494-497.
21. Speicher, M.R., Ballard, S.G. and Ward, D.C. (1996). Nature Genetics12: 368-375.
22. Stack, S. (1973). J.Cell Sci. 13: 83-95.
23. Sutton, W.S. (1903). Biol. Bull. 4: 231-251.
24. Taylor, J.H; Woods, P.S. and Hughes, W.L. (1957). Proc. Nat. Acad.Sci. U.S.A.; 43: 122-128.
25. Watson, J.D.and Crick, F.H.C. (1953). Nature (Lond.) 171: 737-738.
26. Bridges, C.B. (1916). Genetics 1:(1-52): 107-163.
27. Hecht, N.B. and Stern, H. (1971) Exp. Cell. Res. 69: 1-10.
28. Bateson, W. and Punnett, R.C. (1905-08) Rep. Evol. Cttee Roy. Soc. 2-4.
29. Ved Brat, S.(1966). Chromosomes Today 1: 31-40.
30. Zhong, X.B; de Jong, J.H. and Zabel, P. (1996). Chromosome Research 4: 24-28.
31. Ito, M; Hotta, Y. and Stern, H.(1967). Develop. Biol. 6: 54-77.
32. Sturtevant. A.H. (1916). J. Expl. Zool. 14: 41-59.

3

Structure Variations of Chromosomes

The chromosomes provide physical basis by which genetical stability and continuity of individuals and populations are maintained in each generation. Chromosomes also play an important role to generate variation which is an essential component of evolutionary process. Although crossing over during meiosis is responsible for generation of variability, but structural changes in chromosome too are important for causing variation and evolution, since genetic and cytological analysis of individuals of related species or isolated populations within the species have shown that they may differ by various kinds of changes in their chromosomal structure. The structural changes usually may result from spontaneous chromosome breakage. But under normal condition/s such breakage is an rare event and its frequency is dependent upon environmental conditions, i.e., temperature, various high energy radiations (e.g. X, λ and β-rays) and some chemicals. Chromosome breakage is reported to be higher in some hybrids and in presence of particular genes. Certain chromosomes and chromosomal segments (centromeric and other heterochromatic regions) are more likely to undergo breakage. But stability and continuity of structural changes in the chromosomes depend on the behavior of the broken ends. There is no change if broken end rejoins at the same chromosome but a change may occur if it rejoins with the end of other chromosome or if remains unattached. In case of a pair of homologous chromosomes if one chromosome has any structural change and other one is normal we customarily call it a **structural hybrid**.

Deficiencies

The loss of a chromosomal segment from a standard chromosome or of one or more chromosomes is known as deficiency.The terms deletion and deficiency are used synonymously, however, a large deletion in a eukayotic chromosome is referred as deficiency. A broken chromosomal segment devoid of centromere behaves irregularly at spindle and ultimately be lost from the chromosome complement.But in case of plants which have diffuse centromeres,the broken segment can behave normally on the spindle and not necessarily be lost.

Types of deficiency

1. **Terminnal**: In case of terminal deficiency, the lost chromosome segment is terminal one (telomere) and it results from single break in the chromosome (Fig 28A) and pairing between normal and deficient chromosome at pachytene will occur as given in Fig. 28B.

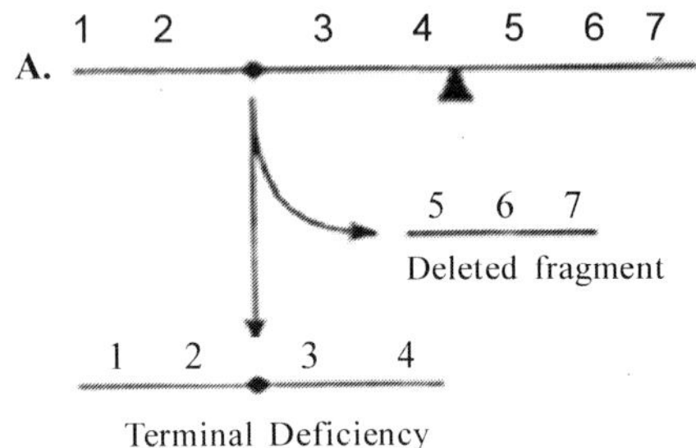

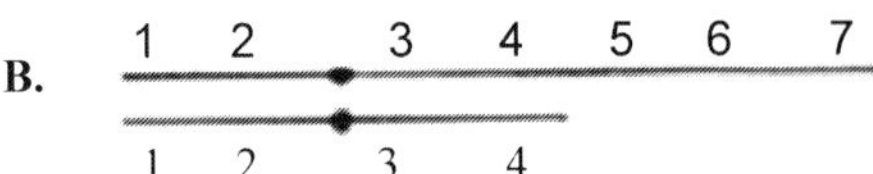

Fig. 28A and 28B: Show origin of terminal deficiency and pairing of a normal and deficient chromosome, respectively.

2. **Intercalary or interstitial deficiency**: In intercalary deficiency,the chromosomal segment between telomere and centromere is deleted or lost followed by rejoining of the raw ends (Fig.29). Intercalary deficiency originates when there are two breaks in a chrosomal arm.Pairing at pachytene between a normal and deficient chromosome forms a loop in normal chromosome at pachytene (Fig. 29C).

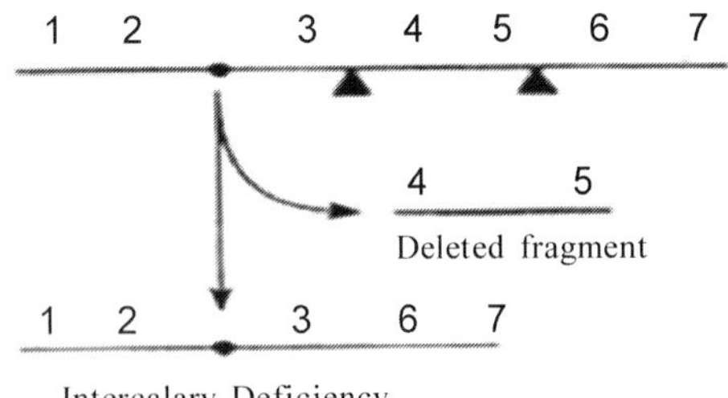

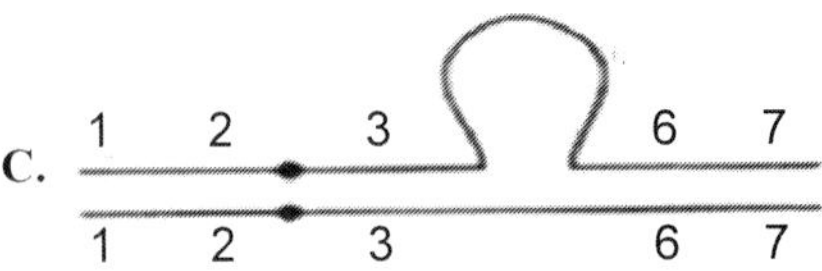

Fig. 29: Loop formation during pairing of normal and deficient chromosome of a homologous pair.

Cytological detection

The cytological detection of small deficiency is difficult or impossible but a long one can be detected directly in suitable chromosomes such as those in drosophila salivary gland in which the absence of bands is evident. In case of corn and tomato, a deficiency heterozygote at pachytene shows a loop. The close pairing of chromosomes during meiotic prophase with loop in deficiency heterozygote, the location of loop indicates the location of missing segment of a particular chromosome. In case of terminal deficiency, the normal chromosome of deficiency heterozygote shows an unpaired segment (Fig 28B) or unpaired segment may be observed to have folded back. However, sometimes bivalents carrying deficiency may even fail to pair.

Genetic behavior of deficiency

The absence of deleted genes, located on the deleted fragment of chromosome, is obviously felt in deficiency homozygotes, however it depends upon the importance and number of deleted genes. In case of flowering plants even small deficiency is lethal to both pollen and ovules, hence only normal offspring is produced.The transmission of deficient male gamete during fertilization is hampered in comparison to the normal one since haploid gametopohytic stage is particularly sensitive to eliminate or sieve for such deficiencies. However, the transmission of deficiency occurs from female side since female gametophyte is less independent of other tissues than the male, hence female gametophyte may tolerate deficiencies. So, in maize no deficiecy within the short arm of chromosome were transmitted via pollen while female gametes could function when the distal third of the short arm was deleted.

Non lethal deficiencies may be detected by decreased recombination of the genes on either side of deficiency or by unusual phenotypes produced by them like dramatically altered sepal and petal color in *Oenothera blandina*. In maize deficiency produced effects resembled to known gene mutants such as brown midrib and white seedlings etc. due to loss of relevant genetic information. Such observations established correlation between cytological and genetical data. In plants heterozygous for a particular locus,the loss of chromosome segment carrying dominant allele expressed recessive trait. The phenomenon is termed as **"pseudo dominance"**which is exploited for gene location on a particular chromosome.

In case of drosophila homozygous males for the dominant alleles were crossed to females carrying recessives The dominant homozygous males were x-rayed to create deletions for this dominant marker. The F1 flies that showed recessive character were crossed again to the multiple recessive stock and the larvae were examined cytologically using salivary glands.Deficiecies could be recognized by the fact that one member of the synaped chromosome pair was shorter and had missed certain bands. Hence this chromosome could be pinpointed to have the gene/s to produce the missing bands.

Chromosomne mapping in plants also has been done by exploiting the phenomenon of **pseudominance.** In plants a stock having recessive marker (aa) are crossed as ♀with irradiated pollen from a ♂ plant having dominant marker (AA) which becomes (A—) after one A is deleted due to exposure to irradiation.To create such a deficiency even seed of heterozygous (Aa) can be exposed to irradiation The F1 of such a cross will be (Aa) or (a—). The plant having genetic constitution (a—), will express pseudodominance and can be subjected to cytolocal analysis for specific pairing like loop formation or shortness of a chromosome in a homologous pair to pin point the deficient gene/marker (A) on that chromosome (Fig.30).

♀ aa x ♂ AA-------(pollens X-rayed)

Gametes (a) ↓ (A) and (--)

Progeny A* and (a--) ; (a-- , shows pseudodominance)

Fig.30: Method to locate a gene on a chromosome by exploiting phenomenon of pseudo-dominance.

The dominant progeny (Aa) is rejected and pseudodominant progeny is subjected to pachytene analysis to correlate loop possessing chromosome with missing marker gene (A). The technique of pseudodominance was extensively utilized for location of gene in corn (Stadler, 1933, 35; Mc clintock, 1931, 1933, 1934 and Singleton, 1939) and in tomato (Khus *et al.*, 1964, Rick and Khus 1966; Khus and Rick 1967b, 1968b).

Breeding behavior of deficieincy

Generally the deficiency is lethal to both ovule and pollen and hence the resultant progeny would be normal one. However, Stadler (1933 and 1935) observerved 50% normal pollen and deficiency bearing transmitted through pollen as well as egg in maize. But deficient embryo sac were smaller than the normal ones. When a plant having longer deficiency were selfed, the F_2 population deviated from normal 3:1 ratio. However, when deficiencies were small, they were transmitted through egg and pollen both, and normal 3:1 ratio was obtained.

Duplication

The term duplication refers to gain of a chromosomal segment or extra DNA in addition to the normal genetic complement and hence thus resulting nucleus, cell, tissue or individual is said to be duplicated/ hyperploids. Thus caused segmental (chromosome) or DNA gain leads to chromosome mutation (macromutations) if they involve visible amounts of genetic material (chromosomal segment).

Types of duplication

(A) 1 2 3 4 5 6 7 8 9 Normal chromosome

(B) 1 2 3 4 3 4 5 6 7 8 9 (Tandem duplication)

(C) 1 2 3 4 4 3 5 6 7 8 9 (Reverse tandem)

(D) 1 2 3 4 5 6 7 3 4 8 9 (On the different arm)

(E) 10 11 3 4 12 13 14 15 (On the different chromosome)

(F) 3 4 (As a separate fragment)

Origin of duplication

1. Unequal crossing over

The origin of a duplication was most clearly demonstrated by studies of bar eye (narrow eye) phenotype in drosophila. Individuals with normal eye had a chromosome segment (S) with 6 transverse bands visible on the X chromosome in the salivary gland, while flies,with narrower bar eyes, have fewer facets having duplicated S segment. Amongst the progeny of bar individuals could be found those with the double bar phenotype having three S-segments in the chromosome as well as wild type flies with a single S-segment only. The occurrence of all these types could be explained by the phenomenon of unequal crossing over between homologous but nonallelic sites (Fig.31) in the female by which upto eight duplicated segments may be produced.

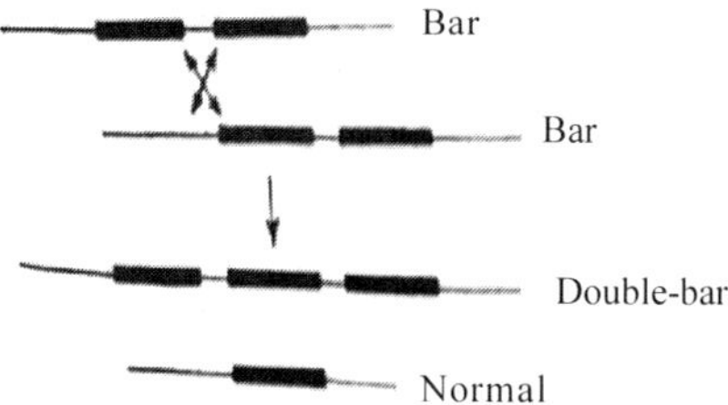

Fig. 31: Phenomenon of unequal crossing over in drosophila leading to formation of double bar and normal eye shape.

2. Breakage-fusion-bridge cycle

This is an another mechanism for production of duplications in corn plant where new or raw ends are formed after chromosomal breakage during meiosis. Such ends may persists into the gametophyte and give rise to the breakage-fusion-bridge cycle. The cycle continues in corn through successive cell divisions in the gametophytes and in endosperm till its completion. Such cycle results into formation of two types of cell phenotypes,i.e., coloured and colourless ones due to duplication or absence of C gene (Fig.32)

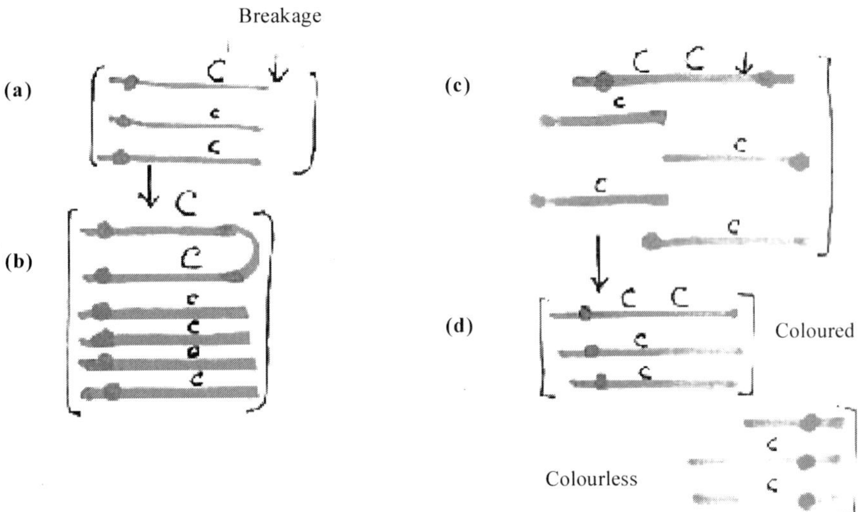

Fig. 32: Duplication of C locus of maize by breakage-fusion- breakage cycle in endosperm (3n). **(a)** broken ends (sister chromatids) in one of the three chromosomes fuse **(b)** at prophase II and produce a anaphase bridge **(c)** which breaks to give various genotypes to have different doses of marker C.; **(d)** producing coloured and colourless sectors.

Pairying of chromosomes of a duplication heterozygote

Tbe pairying between a normal chromosome and its counterpart duplicated homologue results into formation of a loop since that the pairing mate (normal chromosome) is deficient for duplicated segment. Since loops are formed in both deficiency heterozygote as well as duplication heteropzygote (Fig. 33 a&b) but it is rather difficult or impossible to differentiate between loop formed in duplication heterozygote or deficiency heterozygote unless morphological markers are available on the chromosomes. The presence of duplications in normal diploid plant species has come from cytological studies of haploids having duplications which sometime show ring bivalents due to pairing in their duplicated homologous segments otherwise only univalents are expected.

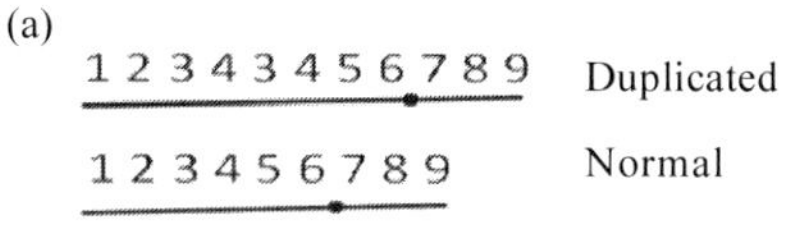

(a)

1 2 3 4 3 4 5 6 7 8 9 Duplicated

1 2 3 4 5 6 7 8 9 Normal

Fig.33a: Duplicated and a normal chromosome.

(b)

1 2 3 4 3 4 5 6 7 8 9

1 2 3 4 5 6 7 8 9

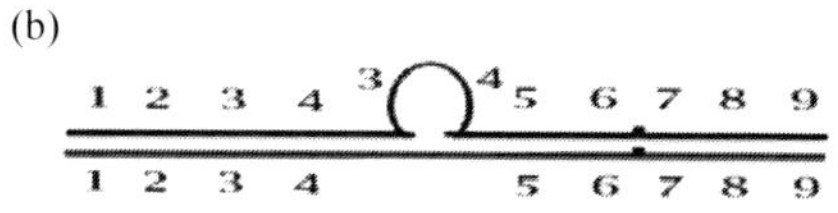

Fig.33b: Pairing between a duplicated and a normal chromosome with loop formation

Consequences of crossing over in a duplication heterozygote

When unequal crossing over occurs in duplication heterozygote, it may lead to various types of consequences as shown in the following figures.

This unequal crossing over will lead to duplication for other segments 5-6 (Fig34a). However, if the duplication is in reverse order as in (34b) it will lead to formation of a dicentric chromosome and an acentric fragment. The consequences of crossing over in figure (34c) will also lead to same results as in 34b. In case the duplication is present on nonhomologous chromosome and crossing over occurs in duplicated region, it would produce two translocated chromosomes (Fig. 34d)

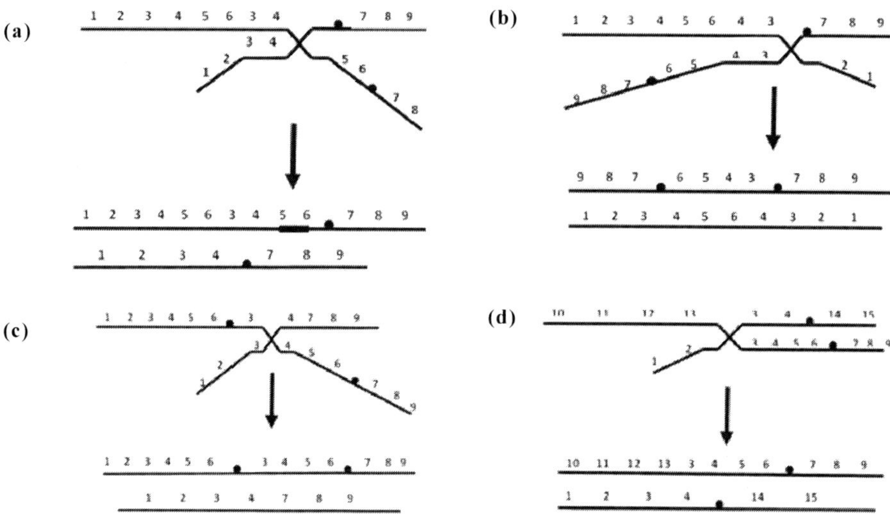

Fig. 34a,b,c and d: Show various consequences of crossing over in on duplication heterozygote

Significance of duplications: Duplications are less deleterious in comparison to deletions and hence they widely occur in natural populations. Pseudoalleles, that is genes which behave as alleles when tested against each other but which are separable by crossing over,originate from duplications. Hence duplication is a mechanism by which new genes can arise and they might undergo further mutation/s to enlarge the range of possible functions available to the organism. Duplication has an advantage that the new variability in the duplicated segment is protected by normally functioning original segment or gene and thus avoiding the usual deleterious effects of most mutations at single loci.

Role of duplication in crop improvement

Hagberg (1962) suggested a procedure to induce directed duplication of a chromosome segment in barley having practical importance. Thus induced duplication of a segment of the short arm of chromosome 6 carried the locus regulating the α-amylase activity responsible for good malting in barley.

Inversion

When two breaks in a chromosome are followed by reunion of this segment, after turning at 180^0 causing reversal of the linear chromosomal segment, was termed as **inversion** by Sturtevant. It basically changes the linear order or sequence of linked genes on the affected chromosome.

The inversion is divided into two types i.e., **paracentric and pericentric** inversion

1. **Paracentric**. In paracentric inversion two breaks are there in the same chromosome arm excluding the centromere. To memorize it, it must be kept in mind that in word para contains only one vowel (a) which may denote a single arm of a chromosome (Fig.35).

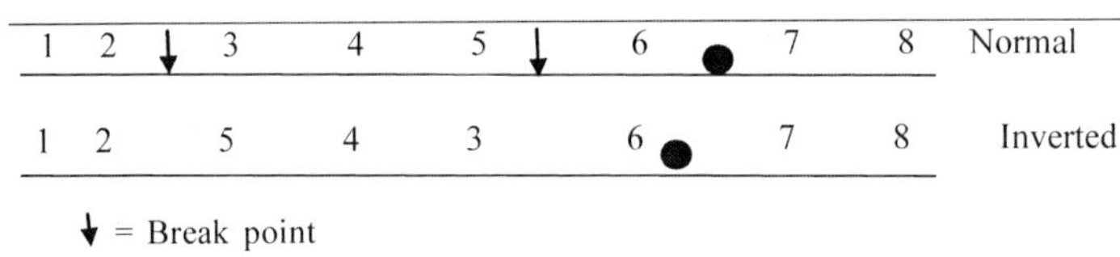

Fig. 35: A normal chromosome with breakages and its inverted version.

2. **Pericentric**. In pericentric inversion two breaks occur in both arms of a chromosome encompassing centromere within these breaks.The breaks in pericentric inversion may be symmetrical or asymmetrical in respect of centromere.Since word peri contains two vowels (e and i) so pericentric inversion can be memorized to have involved two arms of a chromosome Fig.36.

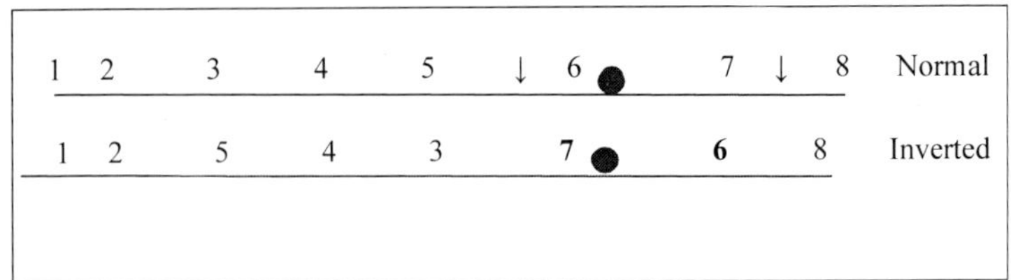

Fig. 36: A normal chromosome with breakages in both arms and its inverted counter part below.

An inversion reshuffles the arrangement of the genetic material in a chromosome rather than its amount. However, both types of inversions differ in their effects on morphology of chromosome and in their genetical consequences. The cytological pairing configurations in heterozygotes of both inversions are similar and to bring homologous parts of normal and inverted chromosome together a "reverse loop " is formed (Fig. 37).

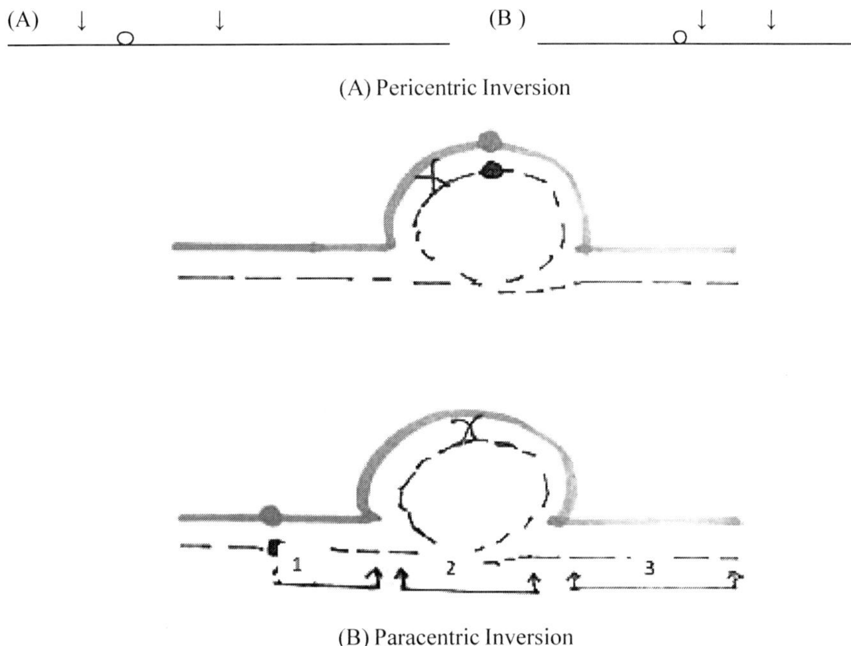

(A) Pericentric Inversion

(B) Paracentric Inversion

1=proximal; 2=Inverted segment; 3=distal

Fig. 37A: The inverted segment is between the arrows (↓) indicating the two break points In the pericentric type, the breaks are around the centromere **(A).** In the paracentric type, both breaks are in one arm of chromosome **(B)** and a typical " reverse loop" pairing of homologous parts in a chromosome pair of a inversion heterozygote. In the paracentric configuration, the proximal segment is the region between the inversion and the centromere. It is referred to as the interstitial segment.

Paracentric inversion in relation to crossing over and fertility

The paracentric inversion is confined to a single arm of a chromosome and its cytological detection can be made by meiotic study.The inversion homozygotes show normal chromosome pairing but inversion heterozygotes, as a earlier stated, form a loop at pachytene and crossing over in inverted segment or loop further determines constitution of resultant chromosomes.In absence of crossing over within inverted region, the chromosomes segregate normally.But when crossing over occurs in inverted region a dicentric chromosome is produced which forms a characteristic bridge and an acentric fragment (Fig. 38). The acentric fragment is unable to move towards the pole due to absence of centromere and hence lost from the dyad and subsequent tetrad. The dicentric chromatid bridge is usually broken by spindle tension or developing cell walls and thus severed chromatid bridge gives rise gametes having duplication or deficiency for some genes whch usually are inviable but those gametes which receive either parental inverted or normal chromosome are viable which are approximately 50 %, in several species The expected reduction in fertility of inversion heterozygotes is not shown on the ♀side where chromatid bridge does not break and forms a chromatid tie and persists in to the second meiotic division. This has the effect of retaining the cross over strands in non functional middle megaspore and thus terminal cell of the linear tetrad which develops into an ovule has high chance to receive one of the parental arrangements (Fig. 39).

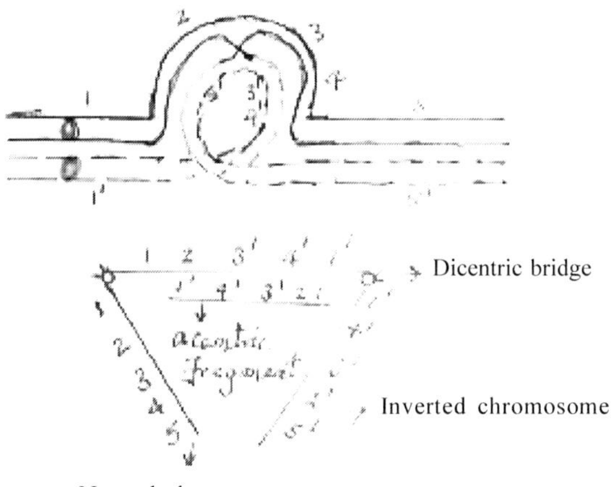

Fig. 38: Crossing over in paracentric inversion which leads to formation of bridge and fragment at anaphase I which persists upto anaphase II.

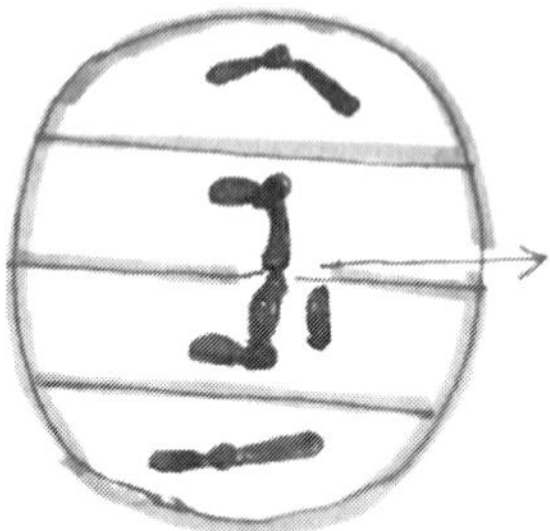

Fig. 39: Chromatid tie at the anaphase I in megaspore mother cell. The cross over products having deficiency and duplication (fig.38) are trapped in middle non-functional megaspores.The gametophyte develops from outer most (terminal) megaspore having normal genetic content.

However, if two or more cross over occur in inversion heterozygotes of paracentric inversion several possibilities do exit (Fig.40). Double cross overs within inversion loop will produce normal gametes if they involve only two chromatids (B+D), 50 % normal gametes if they involve three chromatids (B+C) and no normal gamete if they involve all four chromatids (B+E). If a crossing over inside the inversion loop is associated with one between inversion and centromere (A), it may lead to further complications. The figure 40 and Table1 summarize the all such possibilities.

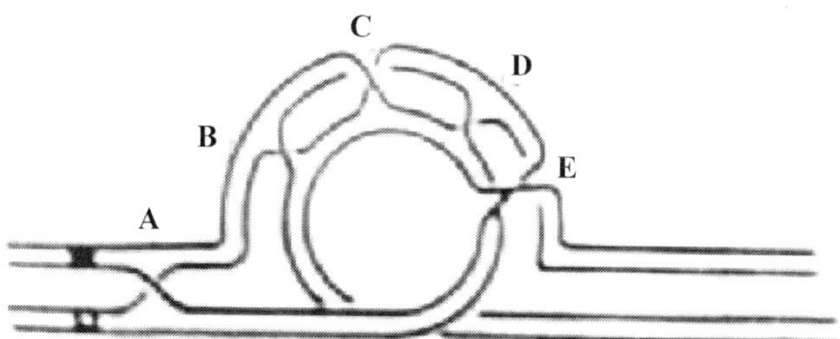

Fig. 40 & Table 1: Consequences of crossing over in paracentric inversion heterozygote and summarized results of crossing over in a paracentric inversion heterozygote.

Crossing over at	Chromosomal configuration	% dupl../def.products
B+C	1 bridge+1fragment at AI	50
B+D	No formation of bridges /fragments	0
B+E	2bridges+2 fragments at A I	100
A+B	Loopchromatid+fragment at A I and 1 bridge+fragment at AII	50

The occurrence of bridges and fragments at anaphase is considered as an indication of paracentric inversion heterozygosity. However, crossing over within the inversion loop in all spore mother cells should produce a single bridge and a fragment of uniform size related to length of the inversion.The occurrence of chromatid bridges at anaphase II is low and inbreeding further reduces the frequency of bridge and fragment eventually leading to reduction of the number of inversion heterozygotes.The bridges and acentric fragments may also arise due to another mechanism of breakage and reunion of paired chromosomes during meiosis. Such spontaneous breaks seems to be localized in the same region as the chiasmata. These are known as U type exchanges and these U exchanges may involve non- sister (NSU) or sister chromatids (SU) in a bivalent and can result from breakage of chromatids during mitotic prophase and pachytene–diakinesis in meiosis. The chromatid exchanges may result in the production of bridge with or without fragments at first and /or second anaphase depending upon the presence and location of chiasmata

The NSU exchanges give a bridge and/or fragment at anaphase-I and these bridges and fragments can persist into anaphase II as a frequent feature contrary to the bridges and fragments of inversion.

Pericentric Inversion

In pericentric inversion, if the two breaks occur from equal distance from centromere , the gross chromosome morphology remains without any gross change as in case of paracentric inversion. However, if these two breaks occur at unequal length from centromere, the relative arm length of the chromosome arms and their arm ratio will change.The pericentric inversion heterozygotes also form a loop at pachytene for homologous segments pairing and a crossing over within the inversion loop also leads to produce normal and duplication - deficient products in 1:1 ratio.But in case of pericentric inversion all the cross over products are centromeric and hence they do not lead to form anaphase bridges and fragments. The crossing over occurring at various locations within inverted segment lead to form balanced and unbalalaced products without bridges and fragments (Fig. 41).

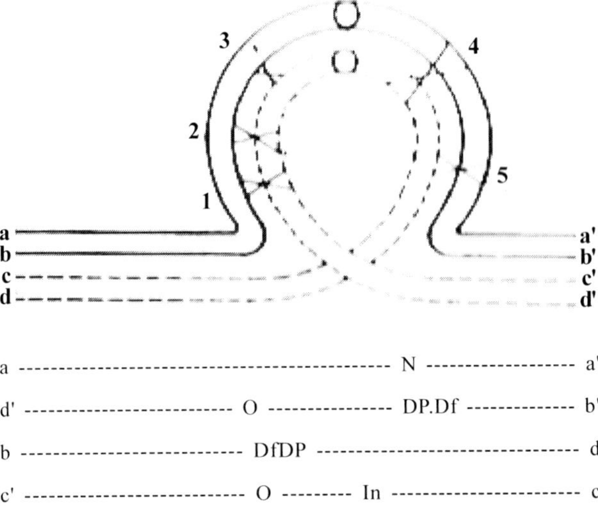

Fig 41. Table 2: Products of a single cross over at position 3 in inverted segment of pericentric inversion. The chromatids a——a' and c—c' are normal and rest of the two carry duplication-deficiency Df-Dp). The crossing over at other locations and their products are given in Table 2 - The double cross overs at various positions within the inversion loop produce balanced and unbalalanced gamemes in the same frequency as in case of paracentric inversion.

Table 2: Cross over products of a pericentric inversion heterozygote

C.O.	Dup.+def chromatids	Chromatids with full complement		% inviable gametes
		Normal	Inverted	
Single CO (3)	2	1	1	50
Doubles 2 strand	-	2	2	0
Doubles 3 strand (3&5)	2	1	1	50
Double 4 strand (3&4)	4	-	-	100

Role of Inversion

1. Presevation of linkage

Both kinds of inversion have the disadvantage for reducing the fertility and suppression of recombination when heterozygous Suppression of recombination is realized from the loss of unbalanced gametes that result from cytologically unchanged recombination. However, the suppression of recombination within inverted region leads to preservation of the linkage between loci in inverted segment at one hand and on the other hand it increases the level of crossing over in other parts of chromosome to create new variation having selective and adaptive advantage.

2. Speciation and adaptation

Sturtevant (1921) invented genetic mapping and published the first evidence of chromosomal inversion that rearrangement of genetic material in chromosome can cause new effects (mutations). He established the genetic map of 3rd chromosome of *Drosophila melanogaster* and *D. simulans* as:

D. melanogaster se st p Dl H ca

D. simulans se st H Dl p ca

These map could be explained on the basis of inverted sequence of underlined three genes, ie, p, Dl and H, obviously a case of paracentric inversion. The reversal of order of these genes r followed by reproductive isolation in these two classes of flies under natural condition thus gave birth to two different species of drosophila, ie, *D melanogastor* & *D simulanse.* White (1978) argued that fixed inversion differences between species are important for post zygotic isolation because of their under dominant (reduced fertility) fitness effects. However, it is improbable that dominant inversions can be selected but Lande (1979) showed through models that population can fix under dominant chromosomal rearrangements that contribute appreciably to hybrid fitness loss. An other alternative hypothesis is that the inversions in fact became fixed because of adaptive differenmces that preexisted at some of those inverted loci.

Translocations or interchanges

Translocations are also structural aberrations in which part of one chromosome is transferred to another non homologous chromosome.The translocations are classified as reciprocal, non reciprocal, internal, jumping, and robertsonian etc.

1. **Reciprocal** : It occurs when two chromosomes are broken at distal part, swapped and rejoined as under:

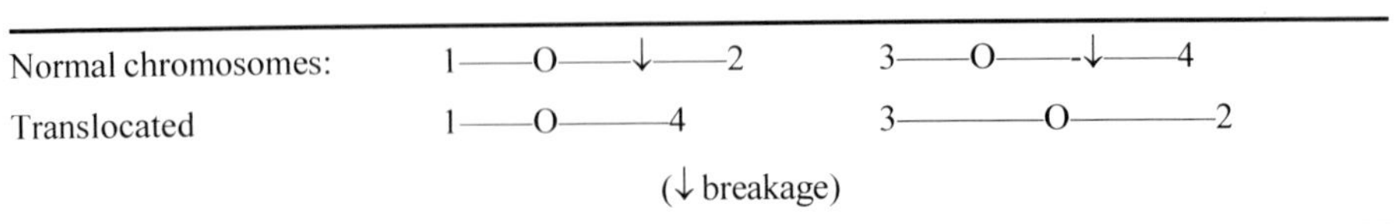

2. **Nonreciprocal** Involves one way transfer of broken chromosomal fragment as:

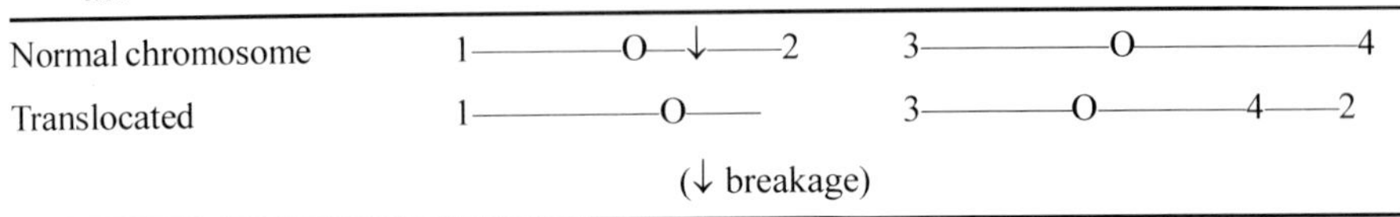

3. **Internal :** It occurs when a broken chromosomal segment is internally inserted within another chromosome.

Normal chromosome	1——O—↓—2	3——O——4
Translocated	1——O—	3——O—2——4

4. **Jumping translocation**: A special kind of terminal translocation where same chromosome segment jumps sequentially to different chromosomes and appears in different terminal positions in different cells

5. **Robertsonian or or whole arm translocation:** It is an extreme form of nonreciprocal translocation, where two acrocentric chromosomes are fused at or near the centromere to generate a compound chromosome which there after segregate as a single unit. But stable centromere like telomeres do not seem able to fuse, hence a centric fusion would require a break along the mid point of each centromere permitting true whole arm interchange. However, there is no clear example that this occurred but most cases of supposed centric fussion appear to involved a Robersonian translocation between two acrocentric non homologous chromosomes.

The breaks in the short arm of one acrocentric and in the long arm of the other, both very close to centromere, followed by interchange will result in a long metacentric and a very short metacentric chromosome,which can be little more than a centromere (Fig.42). Such short chromosomes are usually lost because they to form chaismata in a regular way. If a small chromosome carries little genetic material its loss will not be too deleterious and thus mechanism of reduction of chromosomes operates in nature through Robertsonian translocations. The Robertsonian relationship between original and the compound or derived chromosome complements can be confirmed by meiotic configuration in a heterozygote. In a classical hybrid between *Crepis neglecta* (2n=8) and its derivative *C. fuliginosa* (2n=6), two non-homologous chromosomes of the former species paired with only one chromosome of the latter, indicating that most of the material in the B and C chromosomes of *C. neglecta* has been derived by interchange.

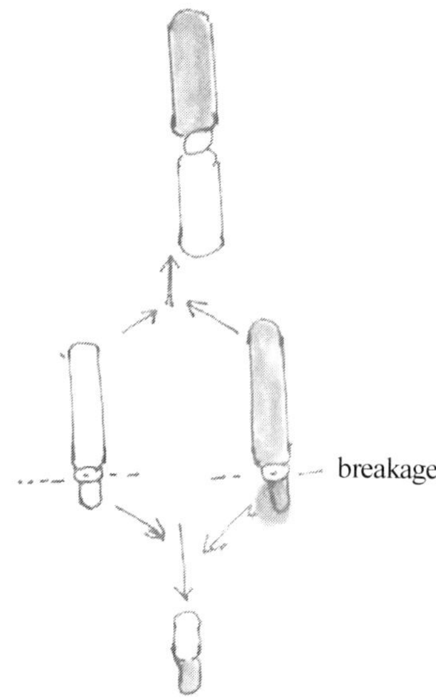

Fig. 42 : Mechanism of Robertsonnian translocation leading to production of a long metacentric and a very short metacentric chromosome (usually lost) from two non-homologous acrocentric chromosomes

Cytological configurations (pachytene) in a interchange heterozygote

In a plant heterozygous for one interchange,involving a single exchange between two non homologous chromosome segments, the pairing at pachytene brings together homologous segments of these chromosomes giving a cross shaped association of four chromosomes (Fig. 43a). If the pairing is strictly homologous the center of cross indicates the breakage–reunion points, but the mechanical effects of torsion during pairing may not permit it to occur exactly. In the pachytene cross it is ususal to distinguish the 'interstitial segments' between the centromere and the points of exchange from the remaining pairing segments of the chromosomes, since crossing over in both the segments has different consequences.

If the chiasmata are formed in each of the four paired arms of the cross then the four chromosomes form a ring at the MI and progressively lower numbers of chiasmata will give a chain of four (Fig. 43b), a chain of three plus a univalent (Fig. 43c), two bivalents (Fig. 43d), One bivalent plus two univalents (Fig. 43e) or four univalents (Fig. 43f). The shape of the ring or chain at diakinesis and metaphase depend upon the localization of chiasmata and the degree to which they termalize, as well as the length of the chromosome arms.

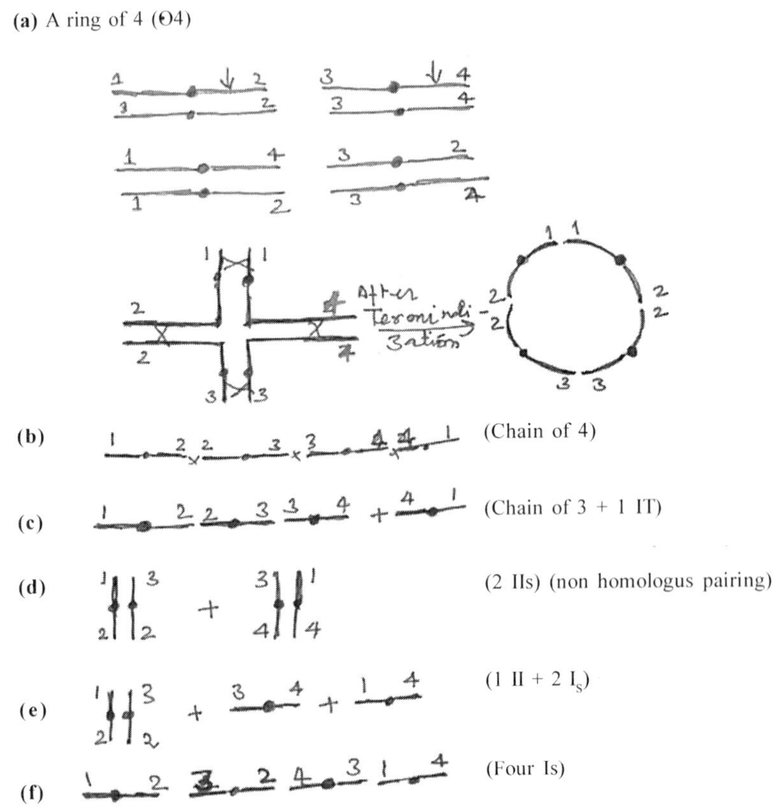

Fig. 43 a-f : Possible configuration of chromosomes in a interchange heterozygote at pachytene.

Orientation and segregation of chromosomes in a interchange heterozygote in relation to gametic sterility.

An interchange heterozygote for one interchange, involving a single exchange of two nonhomologous chromosome segment,the pairing at pachytene brings together homologous chromosome segments to give a cross shaped association of four chromosomes (Fig.44). This cross-shaped structure of four chromosomes can orientate so that either adjacent homologous, adjacent non homologous centromeres may move to the same pole at AI and the resultant gametes are both duplicated and deficient for chromosome segments and are thus usually inviable. However, a twist in the ring compel the adjacent centromeres to move to opposite poles, giving rise to alternate disjunction with full complement chromosome segments, in either the normal or translocated sequences (Fig.44).

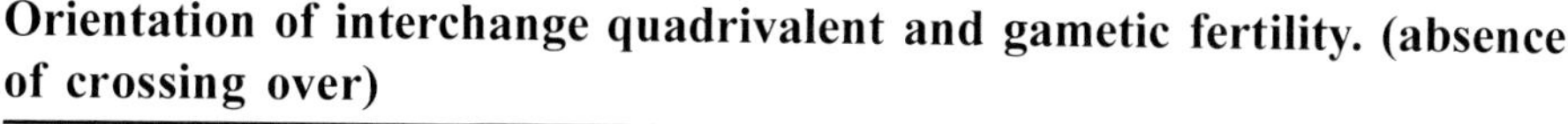

Orientation of interchange quadrivalent and gametic fertility. (absence of crossing over)

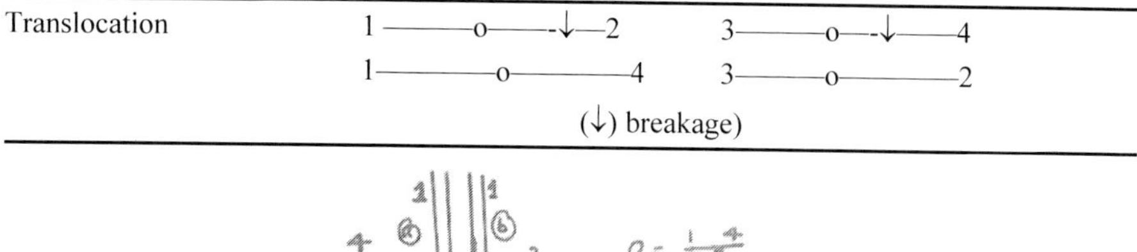

Translocation	1 ———o——-↓—2	3———o—-↓———4
	1————o————4	3———o—————2
	(↓) breakage)	

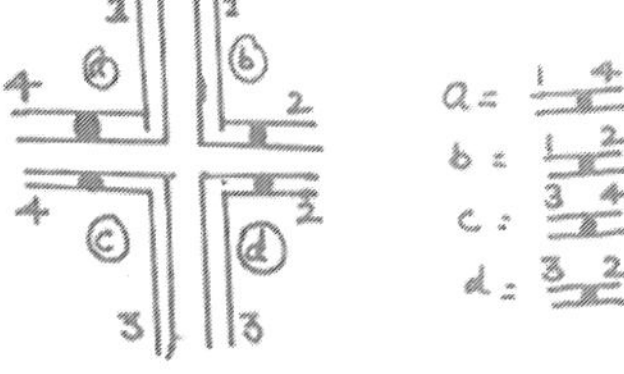

Fig. 44 : Orientation and constitution of chromosomes in an interchange heterozygote

Mode of segregation of centromeres in absence of crossing over

1. Adjacent-I: Homologous centromeres proceed to same pole, ie, (a+c) to one and (b+d) to another

1———O———4 1———O————2

3————O——4 Inviable 3————O———2 Inviable

2. Adjacent II: Non homologous cetromeres proceed to same pole, ie, (a+b) to one and (c+d) to other.

1———O————4 3————O————4

1————O———2 Inviable 3—————O—————2 Inviable

3. Alternate disjunction: Alternate centromeres proceed to same pole, ie, (a+d) to one and (b+c) to other.

1———O———4 1————O———2

3———O———2 Viable 3————O———4 Viable

Thus produced approximately 2/3 gametes are inviable and 1/3 are viable. However, in case of maize and some other plants, only 50% gametes are sterile. Burnhan (1956) suggested that the Open and zigzag rings may be equally frequent and it might be expected if for each open type of orientation at metaphase first, there is an equal chance of alternate orientation or in other words adjacent chromosomes may orient with equal frequency towards the same or opposite poles. However, if crossing over occurs the results are very different. The occurrence of crossing over may at three regions of in a cross shaped structure at pachytene, ie, in interchanged region, in non interchanged arm and in interstitial region. The crossing over in interchanged arm and in non interchanged arm do not affect ultimate constitution of the gametes thus produced. However, crossing over in interstitial region leads to produce sterile gametes in spite of the presence of alternate disjunction (Fig.45). Hence in an interchange heterozygote sterility depends upon frequency of adjacent-I, adjacent-II segregation and crossing over in interstitial segment.

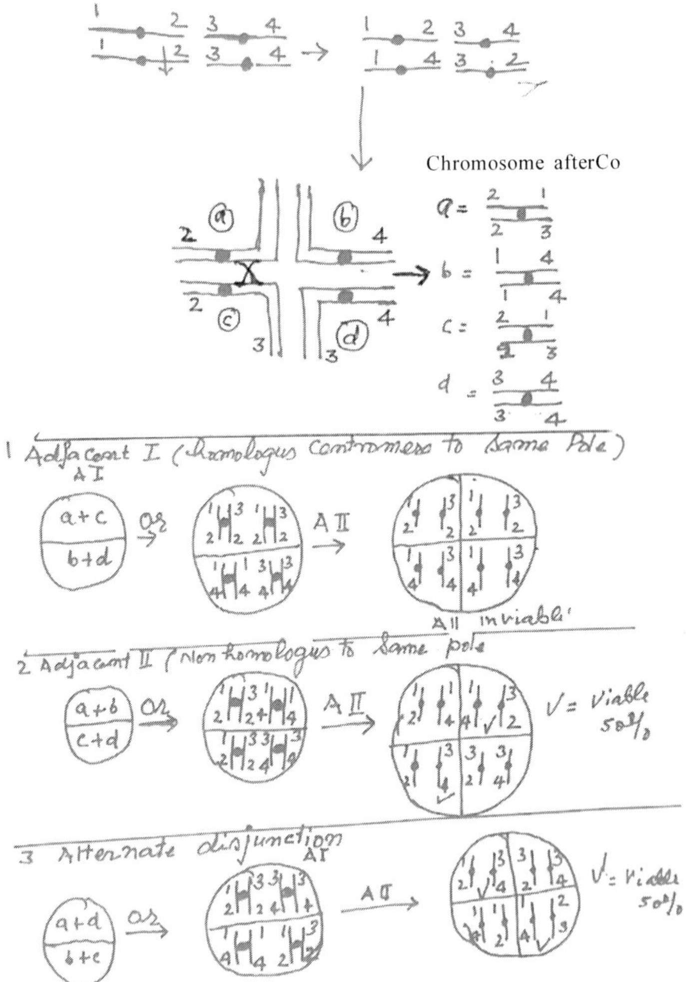

Fig. 45: Segregation of chromosomes in an interchange heterozygote in presence of crossing over in interstitial region.

Formation of multiple translocation rings

If the chromosomes belonging to a pair of homologues individually involved in translocation with two other nonhomologous chromosomes, thus produced translocated chromosomes during meiosis will show a ring of six at pahytene. In the figure 46 , chromosome 3——4 involved in translocation firstly with chromosome 1——2 and secondly with chromosome5—6.. When all these chromosomes after translocation undergo homologous pairing during meiosis , they form a ring of six.

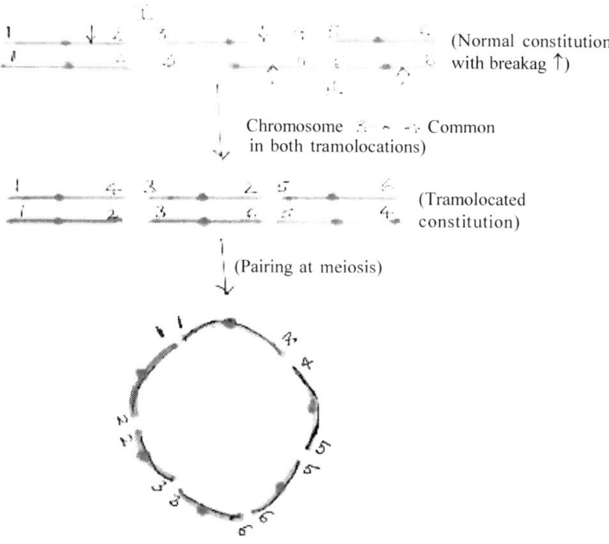

Fig. 46: Mechanism of ring of six (⊙6) formation

Interchanges involving other chromosome pairs likewise can occur to increase configuration and in some cases the whole chromosome complement can be involved, as in species of *Oenothera*, *Rhoeo discolor* and *Paeonia California.* Crossing over in all pairing segments of rings lead to either adjacent or alternate disjunction. However, viable gametes are only produced by alternate disjunction. In case of crossing over in interstitial region, fertility does not exceed more than 50 %. Crossing over in all pairing segments leads to rings having either adjacent or alternate orientation. The alternate disjunction only leads to balanced and viable gametes but in presence of crossing over in differential segment, the fertility does not exceed 50% even under alternate disjunction. Since there is low level of recombination in differential segments in a translocation heterozygote, they tend to be inherited as units undivided by recombination.. Although larger interchange configurations increases the opportunities for orientation , resulting in unbalanced and inviable gametes. But there is evidence from various plants that translocation in equal sized metacentric chromosomes and distal localization of chiasmata and other features of genotype can lead to preponderance of alternate disjunction.

Breeding behavior of an Interchange heterozygote

If a interchange heterozygote is subjected to selfing, its progeny will include normal, translocation heterozygote and translocation homozygote in the ratio of 1:2:1, respectively (Fig.47).

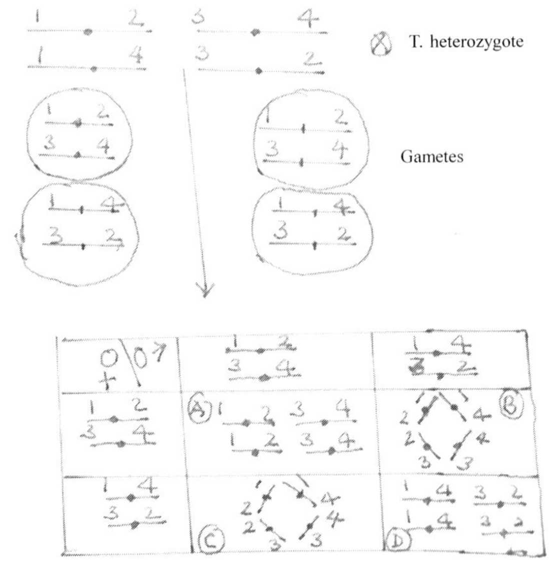

Fig. 47: Breeding behavior of a Translocation heterozygote; under selfing a heterozygote will produce progeny as (A) normal ones, (B&C) *T. heterozygote* and (D) *T. homozygote* in 1:2:1 ratio, respectively.

Based on the breeding behavior interchange heterozygote upon selfing can be distinguished from interchange homozygotes which do not segregate in 1:2:1 ratio upon selfing. However, *Secale cereale*, when selfed gave more heterozygotes than structural homozygotes, suggesting selection against genic homozygosity but in a second interchange no such selective advantasge was observed over its homozygotes and it seemed to be loci and segment specific. Some species, with all their chromosomes involved in reciprocal translocations, maintain permanent structural heterozygosity in spite of autogamous mode of pollination.Such mechanism is well worked out in *Oenothera* where interchange heterozygotes can produce viable gametes under alternate disjunction. Consequently each gamete in *O. lamarkiana* (n=7) consists 7 separate chromosomes which behave as single large linkage group with recombination confined to the pairing ends of each chromosome. Each set of 7 chromosomes which is inherited as a single unit is called a **"Renner complex"**. *Oenothera lamarkiana* has two complexes (**gaudens and velans**) which differ considerably in genetic content.When outcrossed to other species they give very different hybrids. These complexes in homozygous condition prevent the survival of structural homozygotes due to accumulation of deleterious or lethal genes by zygotic lethality (Fig.48).

Similarly gametic lethality can also prevent the formation of homozygotes since each renner complex can only be transmitted via either pollen or egg, thereby giving about 50% abortion of pollen and ovules.

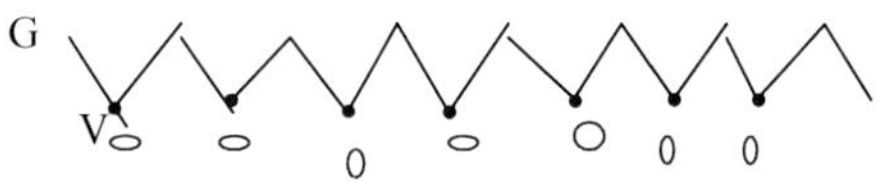

♀/♂	G	V
G	GG	GV
V	GV	VV

GG= Inviable; GV= Viable ; VV= Inviable

Fig. 48: Mechanism of Renner complex in O. lamarkiana. Alternate disjunction results in gametes having 7 chrmosomes of either gaudens (G) or velans (V) set. Homozygous genotypes of thsese complexes are inviable.

The evolution of such breeding systems where complex hybridity is associated with inbreeding has been explained by Cleland (1960) and Darlington and La Caur (1950). Cleland explained that heterozygosity in such cases arose by crossing between outbreeding homozygotes differing by several interchanges and there by stabilization followed the subsequent evolution of autogamy and balanced lethals. However, Darlington holds that complex hybridity arose gradually in response to inbreeding and these two are not mutually exclusive alternatives. Nevertheless, such examples are exception to the view that autogamy usually leads to homozygosity.

Methods of identifying chromosomes involved in translocation

1. **Use of distinctive chromosomal markers**: To identify chromosomes involved in interchange, their cross shaped configuration in a interchange heterozygote can be used provided chromosomes involved have some distinguishing characteristics like presence identifiable centromere,knobs, relative arm lengths and other distinguishing markers in the normal chromosomes are already known. If the position of the center of the cross is variable, the average value of various cross positions is considered. However, this method can only be applied where proper chromosomal spread and stainability of cytological preparations are available.

2. **Comparison of chromosome length and arm lengths of translocation heterozygote, *T.homozygote* with normal plants.**

   To compare these characteristics of chromosomes, the somatic tissue ,ie. Root tips are used. The comparative measurement of chromosome length and other mentioned characteristics from cytological preparations may reveal morpholpogical changes in translocated chromosome/s, provided the pieces exchanged are of different length or the break occurs in distinctive region like satellite.

3. **Use of trisomics**. If primary trisomic set (2n+1) of a species is available, it can be utilized to identify the chropmosome involved in interchange. The crosses are made where trisomics are used as λ♀ and interchanges as ♂. If cytological configuration of F_1 plants show (O 4+1III), it indicates that extra chromosome in trisomic is not involved in the interchange, while a pentavalent or a chain of five indicates its involvement in translocation. The method, however, depends upon the transmission of trisome during meiosis and proper identification of chromosome under investigation. (Fig.49)

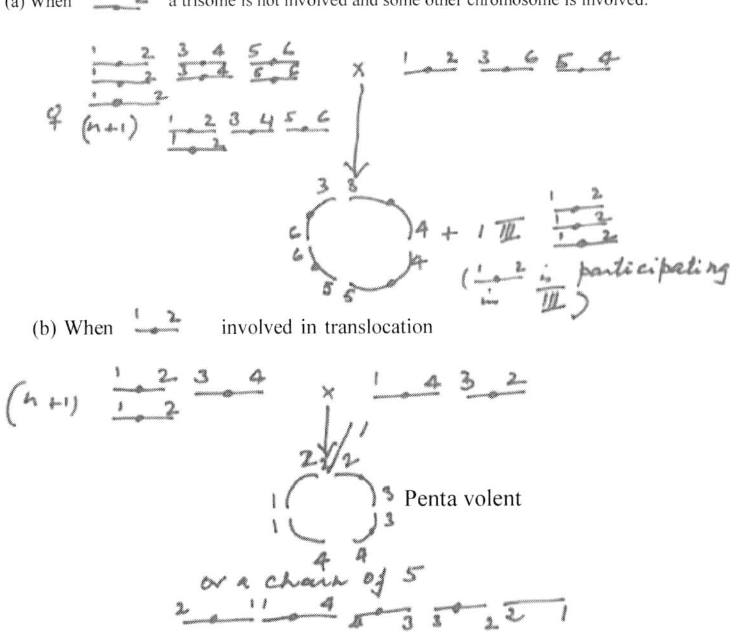

Fig. 49 : Method to ascertain a chromosome involved in a translocation with a trisomic

4. **Inter cross method.** In many plant species ,the chromosomes involved can not be easily identified due to lack of suitable pachytene and prior information on the cytology or genetics of material concerned is available. In such cases inter crossing of unidedentified stock with known translocation stocks (testers) is made and chromosome pairing in meiosis of F_1 hybrid is examined. In such cytological analysis two possibilities exist; one there may be two quadrivalents (2 O 4) or a hexavalent (1 O 6).. The presence of two ring of four (2 O 4) usually means that two interchanges involve different chromosomes and similarly presence of one ring of six (1 O 6) indicates that one interchanged chromosome is common in both interchanges used in the inter cross.

For such a cytological study, a set of tester strains is selected so that a minimum number of crosses required to identify the chromosomes involved in any reciprocal translocation. Burnham *et al.* (1954) used a set of five tester strains in barley (n=7) to identify the chromosomes in a number of reciprocal translocations (Table 3).

Table 2: Critical chromosome associations at diakinesis or metaphase I in hybrids between the tester stocks and unknown translocations to confirm the chromosome involved in translocation heterozygotes for a barley species with n=7 chromosomes

Testrs indicating chromosomes involved in translocation						
	a—b	**b—d**	**c—e**	**e—f**	**c—d**	**translocated chromosomes**
Testers						
a—b						
b—d	1O6					
c—e	2O4	2O4				
e—f	2O4	2O4	1O6			
c—d	2O4	1O6	1O6	2O4		
Unknown						
1	1O6	1O6	1O6	1O6		**b—e**
2	1O6	1O6	2O4	1O6		**b—f**
3	1O6	1O6	2O4	2O4	2O4	**b—g**
4	7II	1O6	2O4	2O4		**a—b**

Table 3: Method to ascertain involved chromosome in a translocation by inter cross of tester stocks.

Role of translocation in relation to plant breeding

1. **Change of linkage and recombination;:** Since during interchanges the chromosome segments are exchanged between non homologous chromosomes,which causes the change in old likage relationships and reestablishes new ones. Burnham (1962) showed that in a translocation $T_{5\text{-}9a}$ (translocation between chromosome 5 and 9) ,waxy (wx) and virescent (v_1) genes were independent, while pr and wx were linked. In normal genetic maps, pr is located on chromosome 5 and wx and v_1 on chromosome 9 assorting indepedently A break in the chromosome 9 occurred between wx and v_1 and that a segment of chromosome 9 carrying wx was exchanged with a segment from chromosome 5 having pr. Thus resulted translocation linked those genes which were previously independent. The interchanges had also affected the level of crossing over. Burnham (1934) showed in the T2—6a heterozygote that crossing over in the Y—Pl region in the long arm

of 6 was only 3.3% where as in the normal stock it was about 28%. The reduction in the pairing was observed and attributed to non homologous or variable pairing.

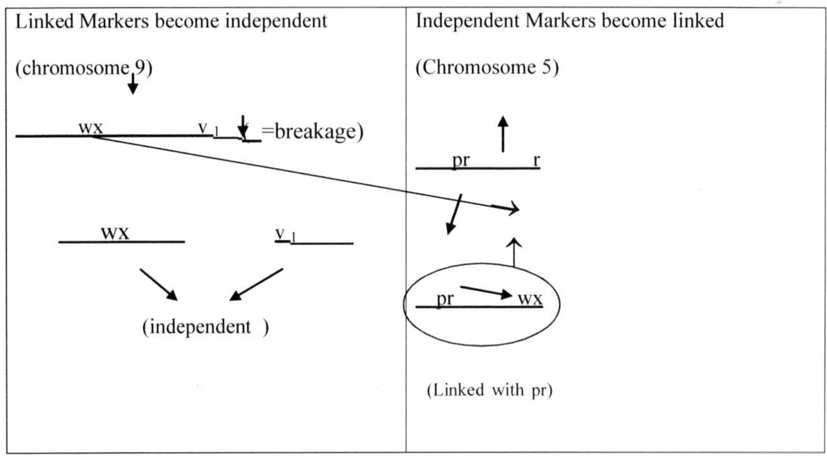

Fig. 50: Impact of translocation on linkage reorientation

Translocation in crop improvement

The pioneer work regarding application of translocation in field of plant breeding was done by Sears (1956) to transfer the leaf rust resistance to bread wheat from *Aegilops umbellulata*. Since seed produced by hexaploid wheat X *Aegilops umbellulata* was inviable, hence an amphiploid between *Triticum dicoccoids* and *Aegilops umbellulata* was first produced to be used as a bridging species which carried *Ae umbellulata* genes to be later on introduced into hexaploid wheat (Chinese Spring). The progeny of Chinese Spring X amphiploid cross was back crossed twice to Chinese Spring and selection at each step produced a plant with 43 ($21^{II} + 1^{I}$) chromosomes. The extra chromosome in the progeny belonged to *Ae. umbellulata* carrying the gene/s for rust resistance alongwith some undesirable genes as well. Since added chromosome did not pair with wheat chromosomes, hence X-rays treatment was used as a substitute for crossing over before meiosis.Pollens from this X-rayed plant were used for pollinating the Chinese spring wheat. The balanced pollen (n=21) would function with or without the possible presence of intercalary translocation. Among 6091 offsprings obtained, 132 were resistant leaf. Out of these 132 plants, 40 had translocation and only one had intercalary translocation "transfer". The plant having intercalary translocation resembled genotypically to Chinese spring except for rust resistance and slight late maturity. The intercalary translocation was later on observed to have involved chromosomes of *Aegilops umbellulata* 6U (Lr 9 gene) and 6BL of wheat (Fig. 51).

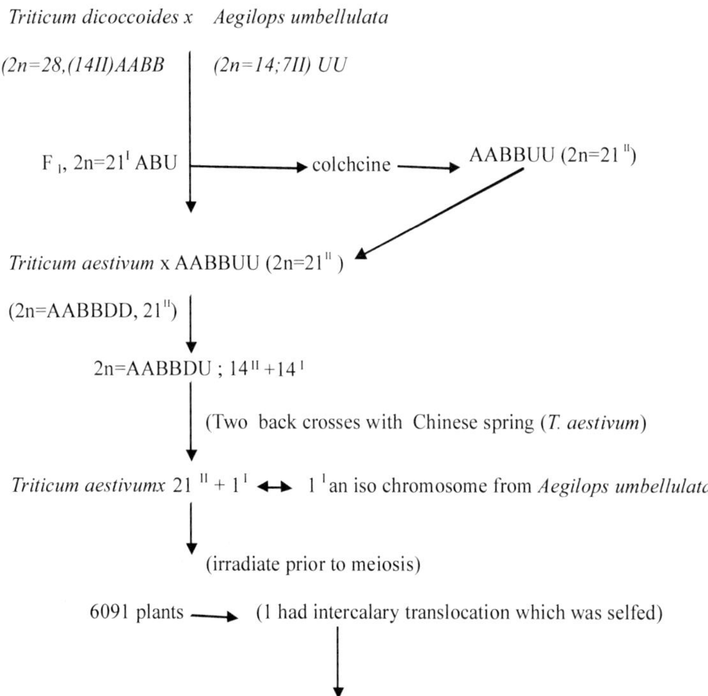

Fig. 51: Scheme of transferring rust resistance from *A.umbellulata* to bread wheat (Sears, 1956).

On the similar pattern of use the irradiation of seeds and plants to produce translocations has been widely exploited in the field of crop improvement. Sharma and Knott (1966 and Knott,1971) irradiated seeds and plants of a substitution line carrying *Agropyron* chromosome. A segment of a chromosome carrying leaf rust resistance was transferred to a wheat variety **Thatcher** and a portion of chromosome 6 el of *Agropyron elongatum* ,carrying stem rust resistant gene Sr26, to chromosome arm of wheat 6A were widely used in improvement of Australian wheats. Another example of a translocation "transec" produced by Driscoll and Jensen (1964) was between 4A of wheat and 2R of rye (4A-2R). It carries resistance for leaf rust and powdery mildew. However, it has not been extensively used due to its low yields which is attributed to its non homologous nature and loss of essential genes of wheat.

Moreover, present day wheat genotypes developed at CIMMYT possess predominantly a translocation/s between wheat and rye. Since breeding efforts concentrated to include rye genetic material in wheat genome by achieving wheat-rye chromosomal translocations 1BL/1RS, 1AL/1RS and rarely a substitutions 1BR. Out of four sources of translocations and substitutions ; two were developed in Germany during 1920-30,one developed in Japan in the 1960s and one developed in USA in the 1970s were progenitors of hundreds of commercial wheat cultivars in different countries of five continents of the world. The source of the 1BL/1RS translocation used by various scientists was either Petkus rye or octaploid (8x) triticale.An analysis of the descendents 172 CIMMYT related cultivars released in 30 countries in 1990s showed that 60% carry in their pedigree 1BL/1RS translocaton. This wheat-rye translocation (IB/IR) ensures high productivity, adaptive possibilities and disease and insect resistance in bread wheat.

Numerical chromosomal changes (Polyploidy)

Chromosome number in somatic cells of flowering plants vary from 4 to264 (Federov, 1969 and Moore and Cave, 1965-1974).The genera and even families may have the same chromosome number in many related groups of plants and species and sometimes populations may also have different chromosome numbers. The difference in chromosome number sometimes may be of one or two chromosome, while in others it may be of multiples of a number, which is equivalent to chromosome numbers found in the gametes of the species and which is considered to comprise the basic chromosome complement (x) or genome of the group. Since meiosis involves halving of the chromosome number, therefore, meiotic products are said to be haploid (n). But cells between zygote formation and next meiotic division consist of two sets of same genome and ,therefore, termed as **diploid** and those consisted of more than two genomes of similar or dissimilar types are called **polyploids**. When chromosome complement of a plant conisted of more than two complete sets of the genome of a single ancestral species; the plant is known as **autopolyploid**. These autopolyploids are further categoriged on the basis of presence of number of identical genomes in their genetic constitution such as (AAA) autotriploid, (AAAA) autotetraploid and so on. The allopolyploids, on the other hand, are consisted of genomes derived from different species brought together by interspecific or wide hybridization. The presence of these different genomes in an allopolyploid determines its nomenclature as an allopolyploid having AABB and AABBDD constitution are called a allotetraploid and allohexaploid and so on, respectively.

Origin of polyploidy

(a) Autopolyploidy

(A) The autoploidy originate either naturally or artificially through following mechanisms:

1. **Syndiploidy** sometimes chromosomes divide more rapidly at mitosis than the cell walls develop between daughter copmplements or spindle fails to develop during anaphase leading to formation of tetraploid sector in an

otherwise diploid plant. If this sector give rise a reproductive tissue, its pollen and ovule will be of diploid constitutution and their fusion would lead to formation of a polyploid.

2. **Restitution nucleus**: Restitution nucleus formed at either meiotic division or during post–meiotic division by the failure of the products to be enclosed by separate nuclear membrane leading to formation of unreduced gametes.

(B) **Artificial production of autopolyploidy by colchicine**

3. **Use of spindle inhibiting drugs (colchicine)** : For production of autopolyploiy in plants, the use of colchine is common and wide. It is an alkaloid extracted from seed and corm of a plant species autumn crocus *(colchicum autumnale*). It inhibits the polymerization of microtubules and thus leading to block the segregation of chromosomes to poles. The methods of its application in plant species may vary but its aqueous solution is usually applied to any meristamatic tissues including seed, seedling, growing shoots and buds of plants.

(b) Allopolyploidy

Allopolyploids are of hybrid origin and their number of chromosomes may have been multiplied before or after the hybridization. Hybrids produced by distantly related species are usually sterile unless their chromosome number is not doubled naturally or artificially. Their sterility is caused by lack of pairing between distinct haploid parental genomes in F_1 which leads to irregular distribution of chromosomes during meiosis to resultant gametes. However, an individual with doubled set of different genomes called an amphiploid (allopolyploid) is generally fertile. The primrose *Primula kewensis* (2n=36) is an example of an amphiploid of *P floribunda* (2n=18) and *P. vertcillata* (2n=18). The F_1 of *P floribunda* (2n=18) and *P. vertcillata* (2n=18) despite of good meiotic pairing is sterile one and so to restore its fertility doubling of chromosome number is achieved by treating its cells with a dilute aqueous solution of colchicine.

Morphological features of polyploids

Autopolyploids especially tetraploids (AAAA) can not always be distinguished by appearance from their diploid (AA) progenitor. However, in some cases they may differ from diploid progenitor in a number of features, such as larger cell size, slower growth rate, delayed flowering and thicker leaf epidermis. In some plants, the length of stomata and the nuclear diameter also increased according to the number of genomes contained. The size of pollens, seeds or fruits is larger in most autopolyploids with reduced fertility. The change in these features is accompanied with the increasing ploidy level, however, there appears to be a threshold beyond which an increase in chromosome number does not increase the size of plant parts.The maximum ploidy level in sugarbeet is reported to be triploid and in corn it is octoploid.

Autotriploids

Autotriploids occasionally occur among the the projeny of diploids as a result of functioning of a unreduced 2n gamete. However, for regular production of triploids, the tetraploids and diploids are crossed. However,in some cases as in barley, the crossing may be difficult to produce the viable seed. Therefore, to overcome such a difficulty **embryo rescue** technique may be employed to obtain F_1 seeds. The results of such crosses were summarized table 4 by Burnham (1962).

Table 4 : Production of triploids by 2n x 4n and 4n x 2n hybridization

Cross	Well filled Hybrid seed	Aborted	% of filled seeds That germinated	No. of plants 3n	2n
2n x 4n	427	20,180	10	48	4
4n x 2n	90	10,170	Most	438	

Cytology of autotriploids: During meiosis an autotriploid may form a trivalent or bivalent plus an univalent. The configurations at diplotene depend upon the number and positions of the chiasmata but only two out of three chromosomes synapse in any chromosome segment. Thus, if chiasmata occur between all homologous chromosomes a multiple types of association may result. An autotriploid can produce a maximum association of three chromosomes in several ways (Fig.52 a-e).

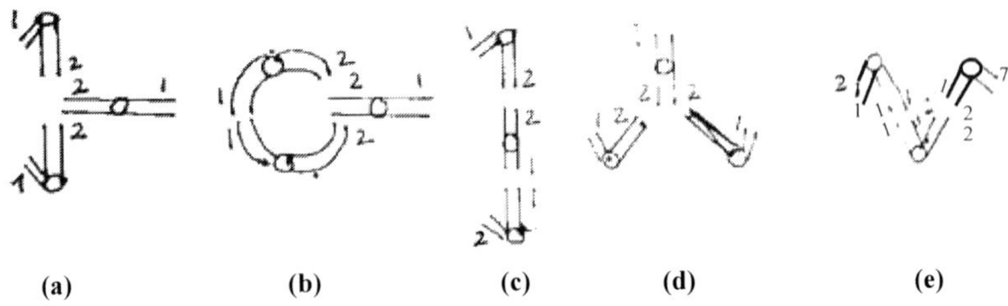

Fig. 52a-e: Orientation of trivalents at metaphase, (a,b) indifferent, (c) linear and (d,e) convergent.

Genetics of autotriploid

Mendelian segregation of alleles in a diploid is determined by the process of meiosis. Since in a diploid only two alleles are there to segregate hence their inheritance is known as disomic inheritance.In autopolyploids the study of the movement and association of the chromosomes is necessary to understand their pattern of inheritace since in autopolyploids more than two alleles are to segregate during meiosis, hence their pattern of inheritance is polysomic. In autopolyploids gametic output depends upon three types of segregations,ie,chromosome segregation, chromatid segregation and maximum equational segregation. The basic difference between these segregations lies in the presence or absence of crossing over between gene and centnromere. The various allelic constitutions of an autotriploid have specific names as: AAA =triplex, AAa=duplex, Aaa=simplex and aaa=nulliplex. The gametic composition, and expected genotypic and phenotypic constitution of an offspring can be determined based upon type of segregation and double reduction.

1. Chromosome segregation
2. Chromatid segregation
3. Maximum equational segregation

The understanding of double reduction is essential for determining the segregation ratios in polyploids since it influences overall outcome of segregation ratios.During meiosis autopolyploids form multivalent chromosome configurations and when crossing over takes place between chromatids of any two chromosomes at a time, the products of crossing over may move to the same pole at AI which is not possible in a bivalent of a diploid. Further more there exists a possibility that the alleles which were previously located on the chromatids of a chromosome before crossing over may also reach to the same gamete after AII. The occurrence of such a phenomenon is known as **double reduction**. However, the occurrence of double reduction requires certain preconditions to be fulfilled as:

1. There must be multivalent formation, ie, trivalent in triploid or quadrivalent in tetraploid since crossing over occurs in paired chromosomes.
2. Crossong over between gene locus and centromere.
3. The two pairs of chromatids resulting from crossing over must pass to the same pole at AI so that sister alleles have the chance of being included in the same gamete.
4. Random separation of the chromatids at AII (Fig. 53).

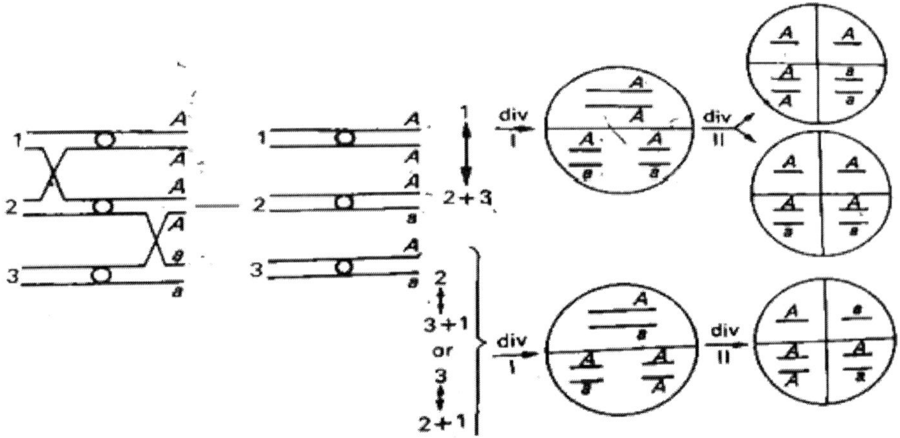

Fig. 53: Mechanism of double reduction ; consequent upon the fulfillment of above mentioned conditions some resultant gametes receive alleles located on sister chromatids as aa gamete.

Chromosome Segregation: If a gene is absolutely linked to the centromere or there is no recombination between the locus and centromere, the segregation of the chromosomes is termed as chromosome segregation. An autotriploid like trisomics produces both n+1 as well as n gametes.The types and proportion of the n+1 gametes can be determined with the help of checker board in a known genotype as given below:

Genotype AAa (duplex)

	A	A	a
A		AA	Aa
A			Aa
a			

n+1 gametes 1AA:2Aa

Aaa Simplex

	A	a	a
A		Aa	Aa
a			aa
a			

2Aa:1aa

The frequency of n gametes, ie, A or a are based on the parental genotype,for example in a duplex AAa and simplexAaa genotypes the ratio will be 2A:1a and 1A:2a, respectively. Furthermore, the ratio of n gametes is brought at par to that of n+1 gametes.based on transmission frequency of n+1 gametes from female side as elucidated in the table 4.

Table 4: Expected gametes and their ratio resulting from chromosome segregation in a autotriploid.

Transmission of n+1 gametes	Genotype	Gametes (n+1)	(n)	Gametic ratio (AA:Aa:A) : (aa:+a)
50% (n+1)	AAa	1AA:2Aa ;	2A:1a	5:1
	Aaa	2Aa:1aa ;	1A:2a	1:1
25% (n+1)	AAa	1AA:2Aa ;	3 (2A:1a) = 6A:3a	3:1
	Aaa	2Aa:1aa;	3 (1A:2a) = 3A:6a	5:7

Breeding behavior of an autotriploid (50% transmission of n+1 gamete)

AAa ♀/0→	2A	1a	**Aaa** ♀/0→	1A	2a
1AA	2AAA	1AAa	2Aa	2AAa	4Aaa
2Aa	4AAa	2Aaa	1aa	1Aaa	2aaa
2A	4AA	2Aa	1A	1AA	2Aa
1a	2Aa	1aa	2a	2Aa	4aa

17A:1a or (17dominant:1 recessive) 12A;6a Or 2 dominant:1 recessive

Chromatid segregation: In chromatid segregation sister alleles are separated into two chromosomes by crossing over and if these chromosomes go to the same pole at anaphase I there is an equational distribution because of both chromosomes have same alleles and furthermore, due to random distribution of sister chromatids at anaphase II some gametes receive sister alleles under process of double reduction (Fig.53). Under random assortment of chromatids in an autotriploid, the formation of n gametes depends on genotype of an autotriploid as in case of chromosome segregation and formation of n+1 gametes can be determined with checker board with 36 squares due to presence of chromatids in an autotriploid.

Duplex (AAa)

	A	A	A	A	a	a
A		AA	AA	AA	Aa	Aa
A			AA	AA	Aa	Aa
A				AA	Aa	Aa
A					Aa	Aa
a						aa
a						

(AAa) 6AA:8Aa:1aa

Simplex (Aaa)

	A	A	a	a	a	a
A		AA	Aa	Aa	Aa	Aa
A			Aa	Aa	Aa	Aa
a				aa	aa	aa
a					aa	aa
a						aa
a						

(Aaa) 1AA:8Aa:6aa

Table 5: Expected gametes and their ratio followed by chromatid segregation

Transmission of (n+1) gametes	Genotype	Gametes (n+1)	Gametes (n)	Gametic ratio (Aa;Aa:A):(aa:a) (Dominant:recessive)
50%	AAa	6AA:8Aa:1aa	10A;5a	4:1
	Aaa	1AA:8Aa:6aa	5A:10a	7:8
25%	AAa	6AA:8Aa:1aa	30A:15a#	11:4
	Aaa	1AA:8Aa:6aa	15A: 30a#	2:3

Adjusted n+1 gametes according to their transmission frequency.

Breeding behavior of AAa and Aaa followed by chromatid segregation

Duplex (AAa)			Simplex (Aaa)		
♀/0→	2A	1a	♀/0→	1A	2a
6AA	12AAA	6AAa	1Aa	1AAa	2Aaa
8Aa	16AAa	8Aaa	8Aa	8Aaa	16Aaa
1aa	2Aa	1aaa	6aa	6Aaa	12aaa
10A	20AA	10Aa	5A	5AA	10Aa
5a	10Aa	5aa	10a	10Aa	20aa

84A:6a or 14A:1a (Dominant :Recessive)   58A:32a or 29A:16a (Dominant :Recessive)

Maximum equational segregation: The chromosome segregation and maximum equational segregation are diametrically opposite to each other in respect to crossing over but the difference between chromatid segregation and maximum equational segregation is negligible. Since in maximum equational segregation there is crossing over between gene and centromere in each chromosome of an autopolyploid leading to maximum frequency of double reduction. The gametic formula for maximum equational segregation can be derived with the help of following Fig. 54.

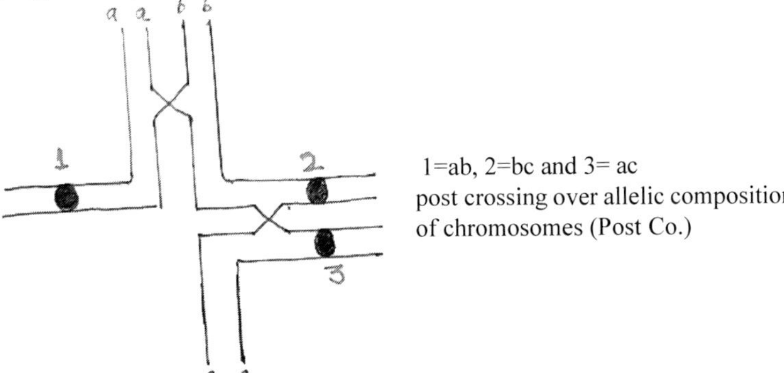

Fig. 54 : Orientation of homologous chromosomes of autotriploid which can be used to ascertain its gametic output

Division	Alternate	Adjacent-I	Adjacent-II
AI	(1+3/2) or (ab+ac/bc) ;	(1+2/3) or (ab+bc/ac) ;	(2+3/1) or (bc+ac/ab)
AII	(n+1/n) (aa+ac+ab+bc/2b+2c)	(ab+ac+bb+bc/2a+2c)	(ab+bc+ac+cc/2a+2b)

= (3ab+3ac+3bc+aa+bb+cc/4a+4b+4c) (There may be three types of orientations hence multiplied by 3)

Hence: 3 (3ab+3ac+3bc+aa+bb+cc/4a+4b+4c)

=**9ab+9ac+9bc+3aa+3bb+3cc+12a+12b+12c** (Utilizing this formula, gametes output by maximum equational segregation can be derived equating alleles of duplex or simplex with a,b and c as follows:

A A a
↑ ↑ ↑
a b c

A a a
↑ ↑ ↑
a b c

Thus the gametic out put of duplex (AAa) will be:

9AA+9Aa+9Aa+3AA+3AA+3aa+12A+12A+12a

=15AA+18Aa+3aa+24A+12a (dividing by common factor 3)

=**5AA+6Aa+1aa+8A+4a** (AA, Aa and aa are n+1 gametes and A and a are n gametes)

By following the same procedure the gametic output in a simplex Aaa will be

1AA+6Aa+5aa+4A +8a

Table 6 : Expected gametes and their ratio produced by maximum equational segregation in an autotriploid

Transmission of (n+1) gametes	**Genotypes**	**Gametes**		**Gametic ratio (AA+Aa+A) : (aa+a)**
		(n+1)	**(n)**	
50%	AAa	5AA:6Aa:1aa 1AA:6Aa:5aa	8A:4a 4A+8a	19:51
25%	AAa	5AA:6Aa:1aa; 1AA:6Aa:5aa ;	3 (8A:4a) =24A:12a 3 (4A+8a) =12A:24a	35:13 19:29

Breeding behavior of an auto triploid followed by maximum equational segregation. The breeding behavior of plants having 25% transmission of n+1 gamete from female side can also be estimated by same methodology.

Duplex (50% transmission of n+1 gamete)

♀/♂	2A	1a
5AA	10AAA	5AAa
6Aa	12AAa	6Aaa
1aa	2Aaa	1aaa
8A	16AA	8Aa
4a	8Aa	4aa

67:5 or 13:1 (dominant :recessive

Simplex (50% transmission of n+1 gamete)

♀/♂	1A	2a
1AA	1AAA	2AAa
6Aa	6AAa	12Aaa
5aa	5Aaa	10aaa
4A	4AA	8Aa
8a	8Aa	16aa

46A: 26a or 23:13 (dominant : recessive)

Autotetraploidy

An autotetraploid contains four sets of homologous chromosomes. They occur when chromosome doubling occurs naturally or artificially in a diploid parent. They have been produced in corn by using pollen from 4n plants (2n x 4n) either from a 4n sector or from unreduced cells in the female parent Bauman (1961). Depending upon chromosome size and chiasma frequency during meiosis the four homologous chromosomes produce five types of chromosomal associations,ie, a **quadrivalent,trivalent +univalent,two bivalents and four univalents**. The trivalents and univalents behave as in triploids. But quadrivalents can orientate in various ways (Fig 55).

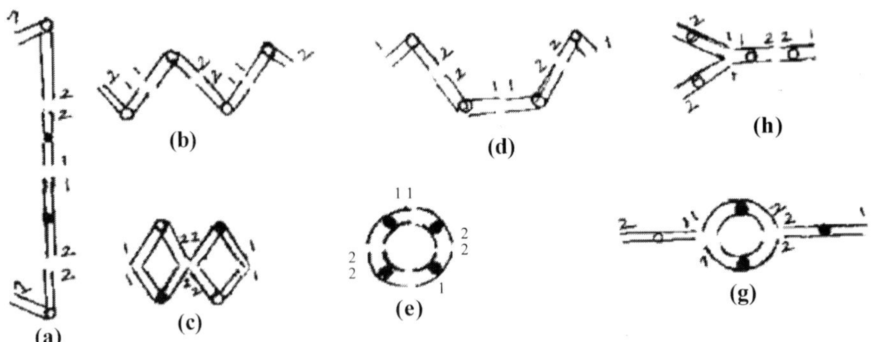

Fig. 55 : various types of orientation of an quadrivalent at metaphase, **(a)** linear; **(b&c)** convergent; (**d&e** parallel and **(f&g)** indifferent; only parallel and convergent give 2: 2 segregation of homologues.

Genetics of autotetraploid: In autotetraploids each gene locus is present four times in somatic nuclei. Hence their inheritance is more complex than diploid ones. A locus with two alleles (A, a) in a diploid lead to formation of three genotypes (AA, Aa and aa) while in a autotetraploid there are five genotypes – AAAA (quadruplex), AAAa (triplex), AAaa duplex , Aaaa (simplex) and nulliplex (aaaa) are formed. A heterogygous diploid produces only two types of gametes,ie, A and a but an autoteraploid produces three types of gametes, ie, AA, Aa and aa (no haploid gamete like triploid) but their frequency depends upon the type of segregation, ie, chromosome segregation, chromatid segregation and maximum equational segregation. The procedure to calculate the gametic types and frequency is almost similar to that of autotriploids but the gametic output under maximum equational segregation is derived with the help of a diagram showing a configuration of four homologues (Fig.57). As stated earlier the level of crossing over determines whether there is chromosome segregation, chromatid segregation or maximum equationaql segregation.

Chromosome segregation: The gametic output in heterozygous genotypes (AAAa, AAaa and Aaaa) under chromosome type of segregation can be derived with the help of a checker board. However, such a derivation of gametic out put is bassed upon orientation of four homologues of an autotetraploid during meiotic division (Fig.56).

Triplex (AAAa)

	A	A	A	a
A		AA	AA	Aa
A			AA	Aa
A				Aa
a				

3AA:3Aa Or 1 AA: 1Aa

AAaa Duplex

	A	A	a	a
A		AA	Aa	Aa
A			Aa	Aa
a				aa
a				

1AA:4 Aa:1aa (Cytological basis given in Fig. 26.

Aaaa Simplex

	A	a	a	a
A		Aa	Aa	Aa
a			aa	aa
a				aa
a				

3Aa:3aa or 1Aa:1aa

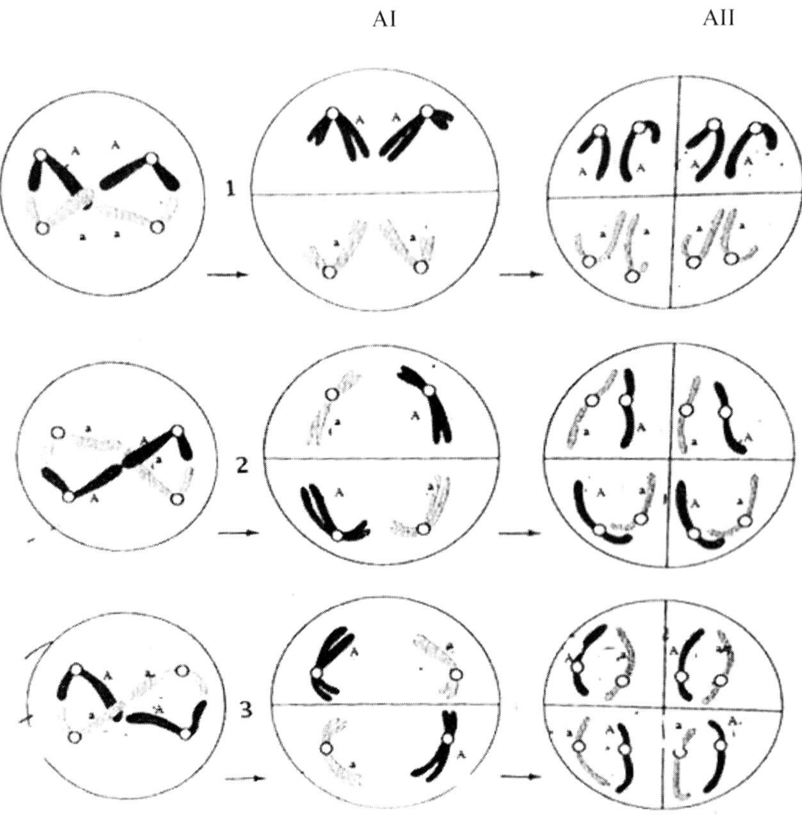

Fig. 56: Cytological basis of 1AA:4Aa:1aa in a duplex AAaa) under chromosome segregation.The first (1) division is reduction division (A with A) and 2 and 3 are equational ones (A with a) and overall gametic output is 2AA:8Aa:2aa or1AA:4Aa:1aa

Gametic output through chromatid segregation in various autotetraploid genotypes, ie, AAAa, AAaa, Aaaa can be determined as follows:

Estimation of genetic output in a triplex, duplex and simplex.

AAAa (Triplex)

	A	**A**	**A**	**A**	**A**	**A**	**a**	**a**
A		AA	AA	AA	AA	AA	Aa	Aa
A			AA	AA	AA	AA	Aa	Aa
A				AA	AA	AA	Aa	Aa
A					AA	AA	Aa	Aa
A						AA	Aa	Aa
A							Aa	Aa
a								aa
a								

Gametes =15AA:12Aa:1aa

AAaa Duplex

	A	**A**	**A**	**A**	**a**	**a**	**a**	**a**
A		AA	AA	AA	Aa	Aa	Aa	Aa
A			AA	AA	Aa	Aa	Aa	Aa
A				AA	Aa	Aa	Aa	Aa
A					Aa	Aa	Aa	Aa
a						aa	aa	aa
a							aa	aa
a								aa
a								

Gametes= 6AA:16Aa:6aa or 3AA:8Aa:3aa

3 Aaaa (Simplex)

	A	**A**	**a**	**a**	**a**	**a**	**a**	**a**
A		AA	Aa	Aa	Aa	Aa	Aa	Aa
A			Aa	Aa	Aa	Aa	Aa	Aa
a				aa	aa	aa	aa	aa
a					aa	aa	aa	aa
a						aa	aa	aa
a							aa	aa
a								aa
a								

Gametes:1AA:12Aa:15aa

Maximum equational segregation: The theoretical maximum frequencies of double reduction in a autotetraplod can be determined with the help of the Figure 57 under three modes of segregation, ie, alternate, adjacent (nonhomologous) and adjacent (homologous)

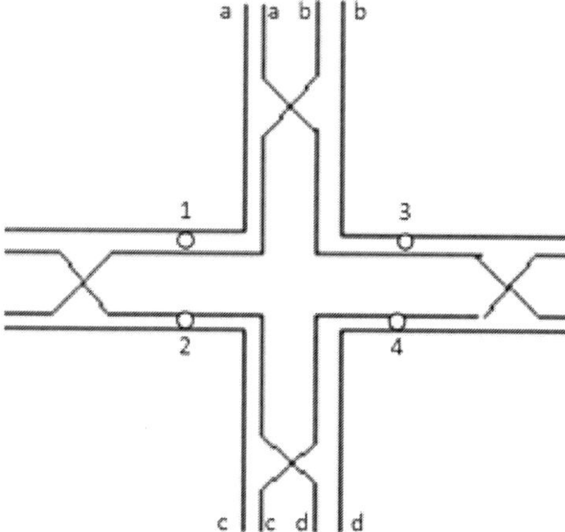

Fig. 57: Gametic out put in an autotetraploid through maximum equational segregation

Post crossing over allelic constitution of chromosomes: 1=ab:2=cd:3=ab:4=cd

Types of segregation

Division I	Alternate	Adjacent (nonhomologous)	Adjacent (homologous)
Centromere	1+4/2+3	1+3/2+4	1+2/3+4
chromatid pairs at the poles	ab,cd/cd,ab	ab,ab/cd,cd	ab,cd/ab/cd
End of division II	ac+ad+bc+bd/ ac+bc+ad+bd	aa+ab+ab+bb/ cc+cd+cd+dd	ac+ad+bc+bd/ ac+ad+bc+bd
Summation	2 (ac+bd+ad+bc)	aa+bb+2ab+cc+dd+2cd	2 (ac+bd+ad+bc)
	= 4ac+4bd+4ad+4bc+2ab+2cd+aa+bb+cc+dd+ 2ac 2bd+2ad+2bc		

Since allelic orientation in the above figure may be of three types, i.e., ab ac or ad, hence overall out put of the crossing over through the same procedure will be:

I	4ac+4bd+4ad+4bc+2ab+2cd+aa+bb+cc+dd
II	4ab+4ad+4bc+4cd+2ac+2bd+aa+bb+cc+dd
III	4ab+4ac+4bd+4cd+2ad+2bc+aa+bb+cc+

Total 10ab+10ac+10ad+10bc+10bd+10cd+3aa+3bb+3cc+3dd

Thus derived gametic series can be employed to determine the gametic genotypes of different heterozygous genotypes, ie, AAAa, AAaa, and Aaaa of autotetraploids by equating the hypothetical alleles with actual allelic constitution as:

a	b	c	d		a	b	c	d		a	b	c	d
↓	↓	↓	↓		↓	↓	↓	↓		↓	↓	↓	↓
A	A	A	a		A	A	a	a		A	a	a	a

So gametic genotypes in various heterozygous will be:

1. Triplex (AAAa)

10AA+10AA+10Aa+10AA+10Aa+10a+3AA+3AA+3AA +3aa or

39AA+30Aa+3aa or 13AA+10Aa+1aa

2. Duplex (AAaa)

10AA+10Aa+10Aa+10Aa+10aa+3AA+3AA+3aa+3aa

16AA+40Aa+16aa = 2AA+5Aa+2aa

3. Simplex (Aaaa)

10Aa+10Aa+10Aa+10aa+10aa+10aa+3AA+3aa+3aa+3aa

=3AA+39Aa+30aa or 1AA+13Aa+10aa

Table 7: Expected gametic ratios in various genotypes of an autoteraploid under chromosome, chromatid and maximum equational segregation

Genotype segregation	Chromosome segregation	A:a	Chromatid segregation	A:a	Maximum equational segregation	A:a
AAAa	1AA:1Aa	(all:0)	15AA:12Aa:1aa	(27:1)	13AA:10Aa:1aa	(23:1)
AAaa	1AA:4Aa:1aa	(5:1)	3AA:8Aa:3aa	(11:3)	2AA:5Aa:2aa	(7:2)
Aaaaa	1Aa:1aa	(1:1)	1AA:12Aa:15aa	(13:15)	1AA:10Aa:13aa	(11:13)

The gametic out put given in the Table 6 can be utilized to estimate the breeding behavior of an autotetraploid. To comprehend the procedure, an actual example regarding breeding behavior of an duplex genotype (AAaa) can be considered. For an example assume an autopolyplod of duplex (AAaa) genotype with the locus of A 50 map units or more from the centromere so that a cross over always occurs in this area and chromatids assort independently. Assuming random assortment of chromatids to the gametes by two's, determine (a) the expected genotypic ratio of the progeny which will result from selfing of this autopolyploid. (b) the increase in the incidence of the heterozygous genotypes compared with selfing of diploids having genotype Aa.

Solution: Hereby the possibility of chromosome segregation (absence of crossing) and extreme crossing over with double reduction can easily be ruled out since random assortment of 8 chromatids or. chromatid segregation is occurring in a duplex (AAaa) which according to Table 87 produced 3AA:8Aa:3aa gametes which will produce the genotypes as per Table 8.

Table 8: (A) Progeny genotypes 9AAAA (quadruplex: 48AAAa (triplex: 82 AAaa (duplex): 48 Aaaa (simplex): 9 aaaa (nulliplex).

♀/0→	3AA	8Aa	3aa
3Aa	9AAAA	24AAAa	9AAaa
8Aa	24AAAa	64AAaa	24Aaaa
3aa	9AAaa	24Aaaa	9aaaa

(B) Selfing of heterozygous diploid (Aa) produces 50 % progeny while selfing of an autotetraploid duplex (AAaa) in the present case produces 178 heterozygous out of 196 progenies. So the heterozygous progenies are 90.8 or 91 % which is 41 % more than the heterozygous progenies produced by selfing of a heterozygous diploid.

Allopolyploidization of autopolyploids: Naturally or artificially produced autopolyploids suffer from several deficiencies like predominantly formation of multivalents during meiosis or disturbed meiosis and low fertility in comparison to diploids. Further more, segregation of more than two alleles in a autopolyploid (polysomic inheritance) due to extensive segregation requires more time to cytologically stabilize it than diploids. Hence in nature autopolyploids under go a possible conversion of their meiotic system partially or completely towards allopolyploidy to evolve reproductive behavior like diploids. The process involved is known as **allopolyploidization**. The genetic causes responsible for allopolyploidization may be a mutation or chromosomal structural change/s leading to differentiation of chromosomes, so that altered chromosomes under go preferential pairing with its own type and not with the original chromosome leading to allopolyploidization. The hybridization of raw polyploids with related diploid or polyploid species is also reported to lead to introgression of fertility genes in them which, upon natural or artificial selection,are responsible for increased fertility and allopolyploidization. Doyle (1979) suggested a scheme of allopolyploidization of autotetraploid maize (ZZZZ) into an allotetraploid (ZZRR), where R genome was derived from original Z genome.

Polyploids in nature and crop improvement

The occurrence of polyploids is very common in nature. cyto-geographical studies have revealed that the proportion of polyploids in over all plant populations approximately doubles from the tropic of cancer towards the Arctic circle. They are more prevalent in harse/extreme climate at higher altitudes. They have considerable economic value since many economic crop plants like cotton, wheat (4x and 6x), coffee, oats, peanut, potato, sugarcane and tobacco etc. are polyploids.Brouwer and Osborn (1999) reported autotetraploid alfalfa to possess winter hardiness an important trait for adaptation. For economic exploitation, autotriploids which are usually sterile and produce seedless fruit are more preferred for human consumption as seedless varieties of watermelon. They are produced by hybridization between 4x (an induced autotetraploid) and 2x genotypes.These triploid melons are high yielding, excellent in flavor, disease resistant and have longer shelf life (Wehner,2008). Moreover, triploids in certain other species are more vigorous than normal diploids as in case of sugar beets *(Beta vulgaris)*. Autotetraploidy has also been used for improving ornamentals for longer blooming duration, to overcome self incompatibility and to be used as bridging species in wide crosses. They are also being used directly as variety in crop plants as tetraploid maize is reported to have 43% more carotenoids pigment and vitamin A activity than diploids.

Allopolyploidy

Allopolyploid cells are composed of two or more sets of non identical genomes of different origins.They may have varying combinations of meiotic behavior and fertility; as complete meiotic stability with reduced fertility, low pairing with complete sterility and complete pairing with low fertility. Their sterility is caused by disharmonious chromosome pairing due to presence of distantly related genomes during meiosis which is usually overcome by doubling of chromosome number. The origin and mechanism of allopolyploidy production can be illustrated by a classical example of an intergeneric cross between *Raphanus sativus* and *Brassica oleracea* to synthesize *Raphanobrassica* made by Georgii Karpechenko in 1928 (Fig. 58).

♀ *Raphanus sativus* (2n=18) x *Brassica oleracea* (2n=18) ♂

(RR 9II) ↓ (BB 9II)

F_1 → RB ($9^{Is}+9^{Is}=18^{Is}$) (sterile)

↓ (chromosome doubling)

RRBB → *Raphanobrassica* (18^{IIs}) fertile

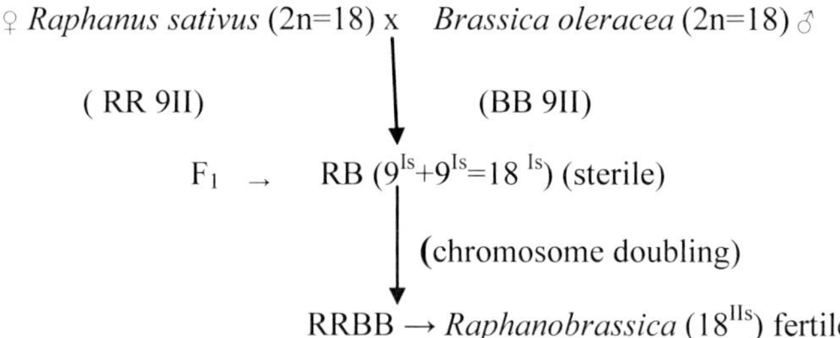

Fig. 58: Origin of allopolyploid *Raphanobrassica* composed of R and B genomes from different ancestral parents, each contributing 9 chromosomes.

The F_1 or hybrid of *Raphanus and Brassica* cross is basically sterile since 9 chromosomes of R genome of *Raphanus sativus* fail to pair with that of 9 chronmosomes of B genome of *Brassica oleracea* due to nonhomology between these genomes. However,occasionally unreduced gametes (RB) were formed which upon selfing led to produce tetraploids in F_2 generation. These tetraploid plants (RRBB), known as **amphidiploids** were fertile and produced 18^{IIs} at meiosis. A large number of economically important crop plants Cotton, wheat (4x & 6x), tobacco and oats are allpolyploids having derived their constituent genomes from distantly related species.In case of origin and evolution of Tobacco *(Nicotiana tabaccum,* **(2n=48; SSTT**); its S genome was reported to be derived from an present day diploid ancestor *Nicotiana sylvestris* (2n=24;SS) and T genome from *N. tomentosiformis* (2n=24; TT), respectively. The contribution of diploid progenitors was established by artificially synthesizing the *N. tabaccum* (Fig. 59A) and study of its meiosis further established homology among its chromosomes and chromosome of its probable progenitors (Fig 59A, 59 B & 59C).

N. tometosiformis (2n=24; TT) x ***N sylvestris*** (2n=24;SS)

(n=12; T) ↓ (n=12S)

$F_1 = 24^{Is}$; 12S+12T) sterile due to non pairing

↓ (chromosome doubling)

2n= 48; 12^{IIs} TT+12^{IIs} SS) (*N. tabaccum*) Fertile

Fig. 59a

N. tabaccum (2n=48; SS+TT) x ***N. sylvestris*** (2n= 24; SS)

(n=24; 12S+12T ↓ (n=12 S)

F_1 = 12 II^s; SS+ 12 I^s ;T

Fig. 59b

N. tabaccum (2n=48; SS+TT) x ***N. tomentosiformis*** (2n= 24; TT)

(n=24; 12S+12T ↓ (n=12 T)

F_1 = 12 II^s of TT+ 12 I^s of S

Fig. 59c

The diploid progenitors, ie, *N. sylvestris* and *N. tomentosiformis* of *N. tabaccum* forms 12IIs each and are fertile. However, their F_1 is sterile having 24 univalents at meiosis without any pairing.When its chromosomes are doubled it becomes fertile and its meiotic division is also normal having 24 bivalents. (Fig. A). When *N. tabaccum* is hybridized with *N. sylvestris* and *N. tomentosiformis,* respectively (Fig. B and C) their F_1 s exhibits 12 bivents+12 univalets establishing the contribution of S and T genome by its diploid progenitors and their genomic homology.

Usually allopolyploids form bivalents as diploids and are genetically stable during meiosis unlike autopolyploids where multivalent and univalents are formed. Nevertheless, the established allopolyoloids like bread wheat are considered to have been derived from hybridization between genetically related but diverse diploid species which have presumably monophyletic origin from an common ancestor having 7 basic chromosome.Hence, the constituent genomes (A, B and D) of wheat have common genes contributing to the same function or phenotype in allopolyploid. The chromosomes of such genomes are called as **homoeologous** (partially homolous). To ascertain homoeology between various genomes, the following criterion are considered:

1. Multivalent formation by chromosomes of different genomes.
2. Frequency of visible mutants in in polyploidy series.
3. Degree of sterility in structural heterozygotes.
4. Duplicate and triplicate gene segregation
5. Compensating effect of specific tetrasomic— nullisomic combinations; as in case of wheat: nullsomic-2D (20) + tetrasomic2B (2) plants are normal ones while null-2D (20) are some what dwarf, leaves short and slightly narrower, male fertile and female sterile. The same pattern was present for plants nullisomic for 20+ tetrasomic for 13.The extra dose of genes in a

tetrasomes compensate the missing vigor of absentee chromosomes of nullisomics. Based upon such a criteria homoeology among A, B and D genomes of wheat was established (Table 9).

Table 9 : Status of homoeology among A,B and D genomes

Genome A	Genome B	Genome D
1A(14)	1B(1)	1D(17)
2A(13)	2B(2)	2D(20)
3A(12)	3B(3)	3D(16)
4A(4)	4B(8)	4D(15)
5A(9)	5B(5)	5D(18)
6A(6)	6B(10)	6D(19)
7A(11)	7B(7)	7D(21)

The homoeologous chromosomes, product of segmental allopolyploidy, would exhibit multivalent formation during meiosis, depending upon the degree of homoeology. Their meiotic behavior and fertility fall in between allopolyploids and autopolyploids. But in case of wheat which is composed of homoeologous genomes, the meiosis is fairly stable and only bivalents are formed. Since it is under control of a genetic mechanism called **diploidization**, a set of processes by which polyploid genomes turns into a diploid one. The basic cause/s responsible for diploidization are genome reorganization involving repeatitive DNA loss, chromosome rearrangements including fusion and fission and complex patterns of gene loss (Dodsworth, Chase and Leitch, 2016) In wheat multivalent formation among homoeologous genomes during meiosis is blocked by a major gene ph1 (pairing homoeologous) located on long arm of chromosome 5 of B genome ($5B^{L}$) in association with some other promoters and suppressors genes (Riley and chapman 1958; Riley, 1966) (Table 8). It is a dominant gene that suppresses pairing of the homoeologues (intergenomic pairing) while allowing that of homologues (intragenomic) to pair causing consequently formation of only bivalents during allopolyploid wheat meiosis.) In absence of 5B (Nuli- 5B), wheat meiosis show multivalent formation involving homologous as well as homoeologous chromosomes. Avivi *et al.* (1982) explained the functioning of 5B gene by a theory of somatic association in wheat based upon the measurement of distances between cytologically marked chromosomsomes at metaphase. Comparison of distances for homologues with those for non-homologues indicated clearly that, within each genome, the homologous chromosomes were significantly closer to one another than non-homologues in somatic nucleus. The close association of homologous chromosomes facilitated quick pairing between homologous chromosomes rather than non- homologues during meiotic division.The components of 5B complex are listed in table 9.

Table 10 : Components of 5B system of wheat

Suppressors	Promoters
1. 2Al, 2BL,2DL	1. 2AS,2BS, 2DS
2 3AS (Ph2) 3BS (Ph2), 3DS (Ph2)	2. 3AL, 3BL, 3DL
3. 4D	3. ———
4. 5BL (Ph1)	4. 5AS, 5AL,5BS,5DS,5DL

However, despite presence of homoeologous genomes in wheat *(Triticum aestivum* (AABBDD; 2n=42) an allo-hexaploid forms only bivalents due to above referred genetical control of pairing during meiosis. The three genomes of wheat are reported to have been derived from three earlier cultivated diploid species. The AA genome of tetraploid and hexaploid wheats are related to A genomes of wild and cultivated einkorn *(T.monoccocom*) and B genome present in tetraploid and hexaploiud wheats is reported to be probably derived from the S genome present in the Sitopsis section of Aegilops *(Ae. speltoids, Ae. longissima, Ae. sharonensis, Ae. searssii, and Ae. bicornis*), with *Aegilops speltoids* being the closest extant species. However, the donor of B genome is still considered controversial. The D genome of *T. aestivum* is considered tohave come from *Ae squarossa* (Kihara,1944).

The genome analysis in wheat, To ascertain the contribution of various diploid species in wheat evolutionary pathyway has been studied by various methods which include (a) study of meiosis in hybrids involving diploid and polyploidy parents, (b) study of karyotypes (Fuelgen stained, Giemsa c-banded or N-banded chromosomes and *in situ* hybridization, (c) immunochemical reactions or protein electrophoresis including study of isozymes and (d) comparison of DNA quantity and quality, etc. including RFLPs from DNA of nuclear, chloroplast and mitochondrial genomes. All the methods have their own merits and demerits but the meiotic study in hybrids is considered a most reliable one (Kimber and Sears, 1987 (Fig. 60 A, B, C and D).

(A) Diploid wheat x tetraploid wheat

2n=14= 7IIs(AA) (n=7;A) ↓ 2n=28=14 IIs (AABB) (n=14;AB)

F_1 hybrid 2n=21(AAB); or $7^{IIs}+7^{Is}$

(B) Tetraploid wheat x hexaploid wheat

2n=28=14 ^{IIs}(AABB)(n=14;AB) ↓ 2n=42 =21 ^{IIs} (AABBDD) (n=21; ABD)

F_1 hybrid 2n=35 (AABBD) (14 ^{IIs}+7^{Is})

(C) diploid wheat x hexaploid wheat

2n=14 = 7 IIs(AA(n=7;A) ↓ 2n=42 =21 ^{IIs} (AABBDD) (n=21;ABD)

F_1 hybrid 2n=28 (AABD) (7 ^{IIS}+14^{Is})

(D) Confirmation of D genome fro m *A. squarossa* by synthesizing artificial hexaploid wheat

Tetraploid wheat (AABB) X *Ae. squarossa* (DD)

2n=28=14 II^S ↓ 2n=14= 7 II^S

F_1 2n=21 (ABD) (sterile)

↓ Chromosome doubling by colchicines

2n= 42 (AABBDD) 21 II^S X 2n=42 (AABBDD)

↓

Synthesized wheat natural wheat

F1 = (21 II) fertile wheat

Fig. 60 A, B, C and D: Meiotic analysis between hybrids of **(A)** diploid and tetraploid wheat **(B)** tetraploid and hexaploid wheat and **(C)** diploid and hexaploid wheat showing presence of diploid wheat genomes in tetraploid and hexaploid wheat and confirmation of D genome.

Synthesis of new allopolyploids: (Triticale -*Triticosecale Wittmack*)

Synthesis of triticales

1. Wheat; A Parent of triticale

1. Wheat, its genomic composition and evolutionary path

Wheat is an important member of the grass family (Graminae) It is classified into three groups according to the number of chromosomes present in their cells. The wheat genome (complete inherited set of wheat genes) is carried by seven chromosomes. As each wheat plant receives at least one complete set of genes from each of its parents and its cells contain at least two sets of chromosomes. Many wheat species have more than two sets, thus they may be diploid, tetraploid and hexaploids. *Triticum aegilopoides* (wild einkorn) and

Triticum monoccocum (einkorn) are diploids (2n=14; AA). *T. diccocum* (emmer) and *T. durum* (hard and macaroni wheat) are tetraploid (2n=28; AABB). The common bread wheat such *as T. aestivum, T. sphaerococcum, T. compactum* which account for most of the wheat grown commercially in present period are hexaploid (2n=42; AABBDD) The evolutionary path of wheat is outlined below. (Fig.61)

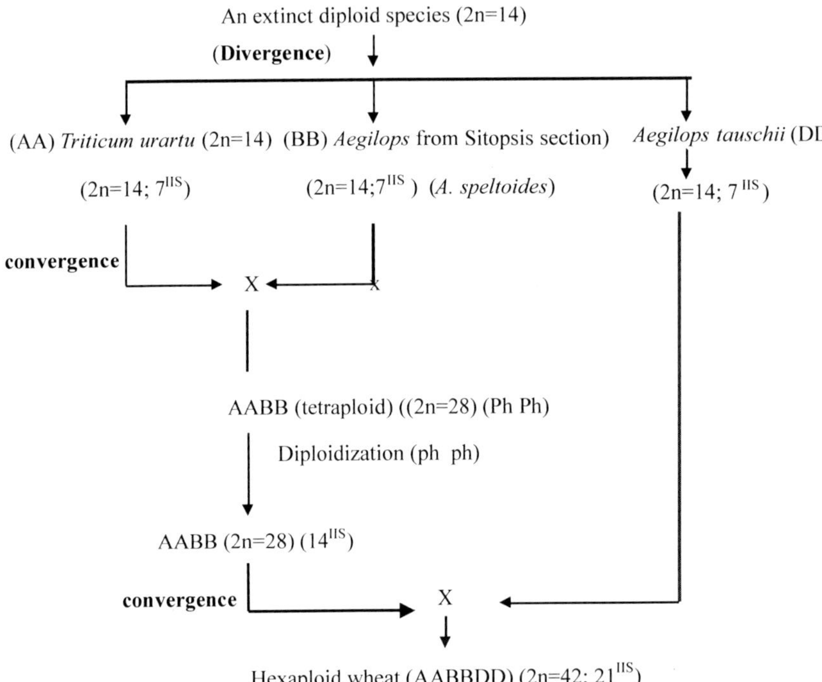

Fig. 61: Evolutionary path of wheat species.

Rye, *Secale cereale*, RR) a second parent of triticale

The rye genome is also carried by seven chromosomes. The earliest known cultivation of rye was in the Balkan peninsula and eastern Europe. Although its many wild species still exist but only one species is recognized for cultivation. It has 14 chromosomes (2n=14;RR).

Triticale (x *'Triticoseale' Wittmack*) Triticale is a new man made cereal, synthesized by hybridizing wheat with rye *(Secale cereale*, 2n=14; RR), with an idea to combine high protein content of wheat with high lysine content of rye. It was also hypothesized that with proper genetic manipulation it would also combine the high yield of wheat with ruggedness of rye (tolerance to biotic and abiotic stresses) along with better adaptability to unfavorable environmental conditions such as cold and dry weather and soils that are light, sandy and acidic. The triticale was first synthesized by a Scottish botanist A. Stephen Wilson in 1875 by pollinating wheat with rye pollen. However, triticale plants thus developed were sterile and could not produce viable offspring. It became a reality only with the development of colchicine induced doubling of chromosomes (Blakeslee and and Avery, 1938) and embryo rescue techniques (Liabach, 1925). However, a systematic triticale research and breeding program could be under taken in North America in 1954at university of Manitoba.

Triticale can be synthesized in two forms: (1) hexaploid triticale and (2) octoploid triticale.If a tetraploid wheat *T.durum* (AABB; 2n=28) and hexaploid wheat (*T. aestivum*; 2n=42; AABBDD) are hybridized with diploid rye *(Secale cereale*; RR; 2n=14), it would result in production of hexaploid triticale (AABBRR; 2n=42) and octaploid triticale ABBDDRR; 2n=56, respectively. The following crosses exhibit production methodology of hexaploid, octaploid and secondary triticales (Fig. 62).

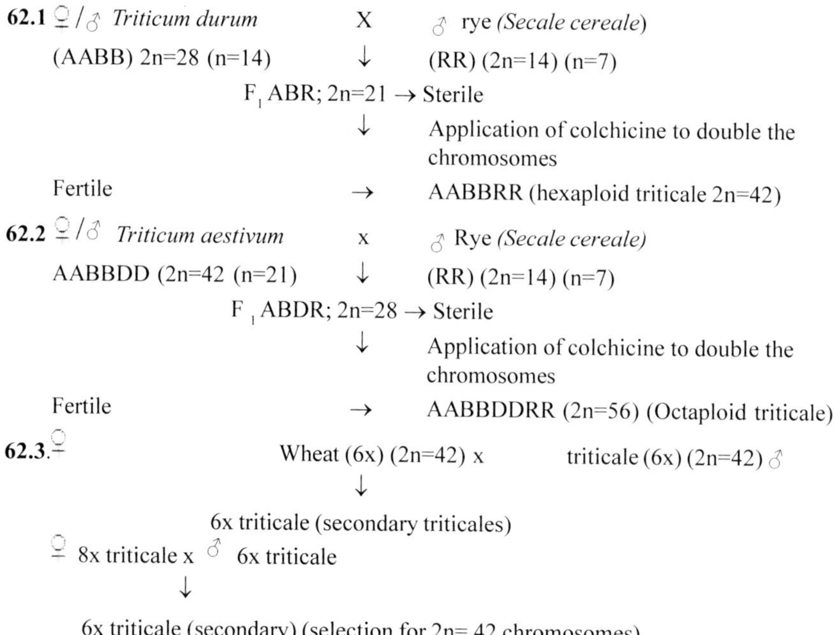

Fig. 62.1, 62.2 and 62.3: Show production methodology of hexaploid, octoploid and secondary hexaploid triticales.

The F_1 generation/s between wheat and rye crosses are sterile since single set of wheat and rye chromosomes are unable to form bivalents during meiosis for purpose of sexual reproduction and thus suffering from sterility. The production of fertile triticale became possible with the application of colchicine (Blakeslee and Avery, 1937) to double the F_1 chromosomes, so that wheat chromosomes could pair with wheat chromosomes to form bivalents and rye with rye. Another important factor in the development of triticale is the development of embryo rescue techniques (Laibach, 1925) since in such wide crosses, the hybrid embryos do not always develop normally, due to genetic incompatibility of the parental species. The chances of survival of under nourished embryos can be increased by removing them 10-21 days after pollination and transfer them to an agar gel containing inorganic salts, other nutrients and sometimes plant growth hormones. Cultured embryos are kept in dark until roots begin to appear and there after allowed to grow in light under constant illumination. When shoots develop, the plants are potted in soil to be treated with colchicine.

Both types of triticales suffer from meiotic instability in terms of univalents at metaphase I, laggards and bridges at anaphase I and micronuclei at quartet but it is octoploid triticale which has higher percentage of these meiotic anomalies. To overcome these deficiencies a third type of triticales were also developed which are known as secondary hexaploid triticales. These are derived from either 6x wheat x 6x triticale or 8x triticale x 6x triticale crosses. These secondary triticales differ from primary hexaploids at nuclear as well as cytoplasmic level. The source of cytoplasm and A and B genomes in secondary tricales is from hexaploid wheats while in primary triticales it was derived from tetraploid wheats. The cytoplasm of hexaploid wheats is reported to be more evolved than cytoplasm from tetraploid wheats and it is reported to be more compatible for hexaploid genotypes of triticale (Larter and Hsam, 1973). And A and B genome of hexaploid wheat are also more evolved than preadapted A and B genomes of tetraploid wheat and thus more compatible for hexaploid genomic triticale constitiution (Thomas and Kaltsikes, 1972) Moreover, at nulear level, the secondary hexaploid triticales have substitution of rye chromosomes by chromosomes from the D genome of wheat (Gustafson and Zillinsky, 1973) which makes them agronomically better performing than primary triticales.

Synthesis of Brassicas

The Brassicaceae comprises of approximately 330 genera and 3700 species and its two genera *(Brassica* and *Raphanus*) are widely grown for edible oils, vegetables, spices, ornamental flowers and as forage crops around the world. The genus *Brassica* consists of three diploid species, namely *B. rapa* L. (2n=20, AA genome), *B. nigra* (L.) Koch (2n=16, BB) and *B. oleracea* L. (2n=18, CC) and their naturally produced allo-tetraploid species *B. napus* L. (2n=38, AACC), *B. juncea* (L.) Czern & Coss (2n=36, AABB) and *B. carinata* A. Braun (2n=34, BBCC). The other genus *Raphanus* contains only one species of agricultural importance, ie, *R. sativus* (2n=18, RR) (Mizushima 1950). In Indian subcontinent, *B. juncea* (2n=36, AABB) commonly known as Indian mustard (Laha, Raya, or Raida), is mainly cultivated for edible oil. The other crop species in this family include *Eruca sativa* L. (rocket salad) *Nasturtium officinale* R. Br. (watercress) *Wasbia japonica* Matsumura (wasabi), *Matthiola incana (L.)* R. Br. (stock) and Erysimum cheiri (L.) Crantz. A number of their wild relatives, on the other hand, have been evaluated as genetic resources for development of new cultivars with biotic and abiotic stress resistance. The famous triangle of U theory establishes close relationship between six important species *B. carinata, B. juncea, B. napus, B. oleracea, B. nigra and B. rapa* and a possibility to synthesize new allo-hexaploid of *Brassica* (Nagaharu U, 1935). The cytology and relationships between six *Brassica* species are well understood and documented (Fig. 63).

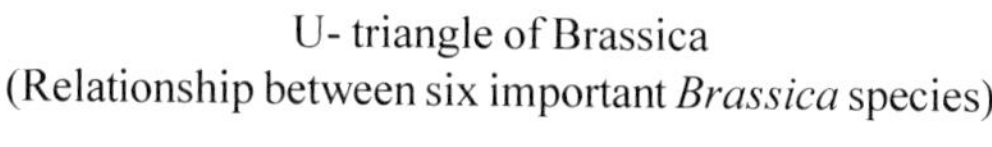

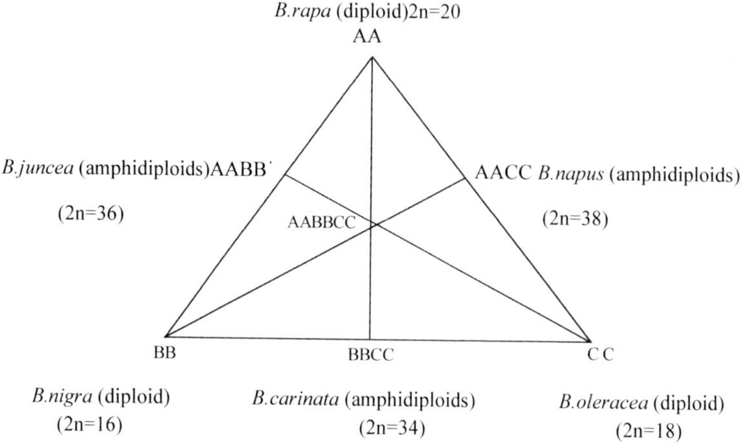

Fig. 63: U-triangle of Brassica and possibility to synthesize new allo-hexaploid (AABBCC)

Since synthesis of *Raphanobrassica*, an intergeneric hybrid between *R. sativus* and *B. oleraceae* by (Kapechenko, 1928), the further hybridizations have been extensively performed to resynthesize the different species of Brassicas to introgress the desirable genes for morpho-physiological variation between cultivated species to develop agronomically potential genotypes (Katiyar *et al.*, 1998). Synthetic amphidiploids in Brassicas have been developed either by doubling the chromosome number of F_1 hybrid plants by colchicine or by fusion of somatic cells. The occurrence of amphidiploids plants are also observed in F_2 generation of amphihaploid F_1 hybrids through fusion of the unreduced F_1 female and male gametes. However, these amphidiploids show low pollen and seed fertility due to unequal chromosome segregation in pollen mother cells in initial generations. With advancement of generations, however, hybrids generally develop high pollen and seed fertility as a result of meiotic stabilization (Kato and Tokumasu, 1976 and Sarashima, 1973).

Synthetic fertile amphidiploids lines derived from hybrids between cultivated crops and their wild relatives have been used as genetic stocks in breeding programs, and as bridging materials for transferring of desirable traits from wild species into cultivated ones. The fertile amphidiploids lines obtained either by chromosome doubling or through somatic hybridization between cultivated crop and wild relatives have led to development of novel crop genotypes as listed below (Table 10).

Table 10 : Novel genotypes, their pedigree and name of the authority for their respective development.

S.No.	Novel genotype//cultivar	Pedigree	Authority
1	Hakuran	*B. compestris x B oleracea*	Namai. 1987
2.	Radicole	*Raphanus sativus x B. oleracea*	-do-
3.	Raparadish	*B. compestris x R. sativus*	-do-
4	Gifu-Green	*B. rapa x B. oleracea*	Takada1985
5	Senpousai 1 &2	*B. rapa x B. oleracea*	Nagano1988
5	Hanakkori	*B. rapa x B. oleracea*	Matsumoto et al. 1997
6	F_1-Hosoda-wase	*B. rapa x B. oleracea*	Takada1906

All the natural Brassicas exist either at diploid or tetraploid levels. Hence, a new approach has recently been initiated to synthesize several hexaploid Brassica populations since plants with higher ploidy levels tend to have more resilience to adapt to harsh climatic conditions than their counterparts with lower ploidy level. In this context the success of hexaploid triticale which has higher yield potential than its parents, ie tetraploid wheat and rye) can be cited (Guesdes-Pinto *et al.*, 1996). The development of hexaploid *Brassica* combining the three Brassica genomes (A,B and C) is under way on the pattern of triticale. Since such *Brassica* hexaploids would widen its genetic base for adaptation to various environmental ranges being benefited from intersubgenomic heterosis (Zou *et al.*, 2010). The artificially developed allo-hexaploids with genomic constitution AABBCC showed promising performances in agronomic traits. The three most important tetraploid Brassica species (*B. napus*, AACC; *B. juncea* AABB & *B. carinata* BBCC) are from the pairwise genome combination of three diploid species (*B. rapa* AA, *B. oleracea* CC *And B. nigra* BB). All the genomes (AABBCC) for constituting the allo-hexaploid have their own specific genes to control various traits of agronomic significance. The allo-hexaploid, as hypothesized, would bring together these genes from constituting genomes to create "Super Brassica". There are three popular approaches to produce hexaploid Brassica hybridizing members as there U triangle, i.e., using three tetraploids as parents by unilateral unreduced gametes, using one tetraploid and one diploid as parents followed by chromosome doubling by colchicine and using three diploids as parents (2X- 2X-2X).

Haploidy, its origin, classification, detection and application in crop improvement

Haploidy

Haploidy is a normal state in gametophytic generation in flowering plants. The natural haploid sporophytes, having gametophytic/ gametic chromosome number and constitution, are called haploids. For example, in case of wheat which has 2n=42 chromosomes, individuals having n=3x=21 chromosomes are haploids. The occurrence of haploids have been observed in almost 100species belonging to 16 families of angiosperms (Kimber and Riley, 1963 and Kasha, 1974). The haploids usually originate through parthenogenetic development of the embryos from the unfertilized reduced gametes sometimes spontaneously. But to produce them in large numbers several other techniques are employed such as by pollinating the irradiated flowers with normal pollens, delayed pollination which induces parthenogenetic development of egg into haploid embryo. The use of alien cytoplasm of *Aegilops* in wheat genotype "Salmon" having 1B//1R translocation induced haploid production (Kihara and Tsunewaki, 1962).

Haploids can also be produced through distant hybridization (interspecific and intergeneric) between distantly related species where egg cell is stimulated to develop even though their gametes are too unharmonious for proper fertilization or fertilization of only one egg cell stimulated the development of another in the same or adjacent embryo sac. In hybrids between *Hordeum vulgare* with *H. bulbosum* and between *Triticum aestivum* and *H. bulbosum*, a high frequency of haploids are produced from the elimination of *H. bulbosum* chromosomes (chromosomal elimination) during meiosis (Kasha, 1974). However, Bennett et al. (1976) confirmed normal double fertilization cytologically in these inter specific crosses but after fertilization a gradual and selective elimination of *H. bulbosum* chromosomes occurred in the nuclei of endosperm as well as embryo cells, resulting into production of haploid embryos. Bennett et al. (1976) examined the possible causes of chromosome elimination and concluded that sudden shortage of essential proteins for developing embryo and endosperm and better ability of vulgare chromosomes to form spindle attachments relative to bulbosum chromosomes could be responsible for bulbosum chromosomal elimination and thus production of haploids.

Recently some most successful techniques have been developed which include production of haploids through culturing of anther, microspore and ovule. Usually culturing of anther is preferred over microspores and for anther culturing, anthers are excised just before, during or immediately after microspore mitosis. The success of culture depend upon parameters like genotype and condition and growth of donor plant, pretreatment, developmental stage of anther/ microspore and culture medium and condition during culture growth. The culture (solid media) consists of nutrient agar media having sucrose, mineral salts and depending on the species, various vitamins, auxins and amino acids. Alternatively anthers can be floated on liquid medium to be taken for to culture isolated pollen grains. Thus cultured anthers/microspore give rise high frequency of haploid callus tissues or embryoids by division of the vegetative cell in microspore and in majority of the cases generative cell has no, or only a vestigial role except in case of *Nicotiana* where it too give rise to embryoids which directly develop into plantlets. Embryoid formation can also be induced from calluses by transferring them to apprppriate regenerative media. The haploid plants have also been produced recently from cultured female gametophytes successfully in some gymnosperms like Zamia, Ephedra and some cycads and crops like barley, wheat and tobacco

The production of haploid plants by culturing needs to be taken utmost care during removal of anthers from filament not to have any injury since wounding of anthers lead to often stimulation of callusing of the anther cell wall tissues leading to generate various levels of ploidy (diploid, triploid, tetraploid or hexaploid) in calluses and so plants derived from such cells would not be of gametophytic origin. Furthermore, plants produced through calluses are prone to chromosomal changes and so needs to be cytologically screened.

Classification of haploids :

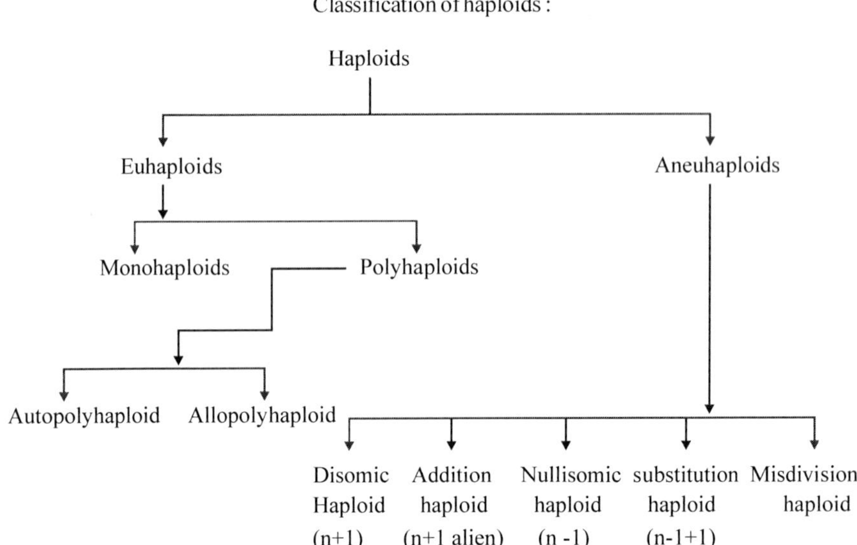

Euhaploids

Euhaploids are those plants which have chromosome number exact multiple of one of the basic numbers of the group and these are further classified into two classes, i.e., monohaploids and polyhaploids.

1. Monohaploidy

Haploids of diploid species having one representative of each chromosome in the basic complement (x) are termed as monohaploid or simply monoploid since they possess a single genome. Because they have a single representative of a homologous pair, hence their meiosis is irregular with no bivalent formation predominantly. However, some pairing does occur as in cases of *Hordeum spontaneum*, *Oryza glaberrima* and *Secale vulgare*, which is attributed to non homologous secondary association with doubtful chiasma formation (John and Lewis, 1965). The equational division for chromatids segregation in meiosis occurs in the second meiotic division. In first division univalents pass to poles without their equational division and since bivalents are not formed, the univalents move to either pole resulting into hypoploid number of chromosomes to either pole. These chromosomes ultimately divide equationally and end up in a deficient set of chromosomes to each of the resultant four microspores with less chance of obtaining complete set of chromosomes essential for viability of a gamete.

2. Polyhaploidy

Haploid individuals originating from auto or allopolyploid species are termed as polyhaploids. They may be dihaploids (AAAA autotetraploid (4X) $\rightarrow$ auto-di-haploid (2X=AA), auto-tri-haploid (6X, AAAAAA $\rightarrow$ auto-tri-haploid (3X=AAA) and so on. Similarly they can also be derived from allopolyploids, as allo-di-haploid will originate from allotetraploid (AABB=4X $\rightarrow$ allo-di-haploid (2X=AB), allo-tri-haploid from allohexaploid (6X=AABBDD $\rightarrow$ (3X=ABD) and so on. The auto –di-haploids (AA) usually have been reported to exhibit frequent bivalent formation during meiotic division in species of *Capsicum, Dactylis, Medicago, Solanum Valerians, Bromus Parthenium and Sorghum,* the last three of which are self-fertile (Kimber and Riley, 1963) (Table 11).

Therefore, these polyhaploids due to normal chromosome pairing and chromosomal segregation produce fertile gametes. A polyhaploid (auto-tri haploid 2n=21) derived from autohexaploid. *Phleum pretense* (2n=42=21^{IIs}) showed bivalents and univalents ($7^{II}+7^{I}$). Despite of the fact that it had hexagenomic ratios and its triploid (2n=63) formed $28^{II}+7^{I}$. Nordensiold (1945) demonstrated these results and suggested that bivalents are only formed in multiples of basic number and rest of chromosomes show themselves as univalents.

Table 11: Somatic and polyhaploid chromosome number and bivalent formation during meiosis in polyhaploids of some polyploids.

Plant species	Somatic chromosome number (2n)		Maximum chromosome pairying at MI (polyhaploid)
	Polyploid	Polyhaploid	
Medicago sativa	32	16	8^{II}
Valeriana sambucifolia	56	28	14^{II}
Parthenium argentatum	72	36	18^{II}
Solanum tuberosum	48	24	12^{II}
Dactylis glomerata	28	14	7^{II}
Bromus inermis	56	28	14^{II}
Sorghu m halepense	40	20	10^{II}

In case of allopolyhaploids (AB) one would usually expect only univalents, since the polyhaploids of true amphidiploids behave exactly like monohaploids, nevertheless presence or absence of some bivalents depends upon the presence or absence of homoeology in the constituting genomes of allopolyhaploid. Different levels of homoeology would produce different results and there could be considerable variation even within the polyhaploids of same parentage; polyhaploid of *Solanum polytrichon* (2n=24), for example, can form 4-12 bivalents (av. 8 bivalents) and sterile gametes. In case of wheat allo-polyhaploid (ABD) bivalents are formed in low frequency since three genomes (ABD) have homoeologous genomes. The polyhaploid of *Triticum aestivum* also has three times as much pairing as that of related monohaploid *Aegilops longissima.*

3. Aneuhaploidy

1. **Disomic haploid** (n+1) : These haploids contain one chromosome in disomic condition, which belongs to the one of the genome of haploid.
2. **Addition haploid** (n+ few alien) : Additional chromospme in the haploid is alien chromosome.
3. **Nullisomic haploid** (n-1). One chromosome is deficient in haploid constitution.
4. **Substitution haploid**: One chromosome of the haploid is substituted by an alien chromosome (2n-1 + 1).
5. **Misdivsion haploid**: These haploids contain either a telo-centric or isochromosome resulting from misdivision of chromosome.

4. Detection of haploids

1. **Morphological features**: In certain cases some morphological features indicate possibility of haploidy, as Mehetre and Thombre (1981) identidfied miniature stature, small leaves and flowers, shorter internodes and anther indehiscence differing significantly in haploids and diploids in case of cotton. Moreover, the haploids were completely sterile to semi-sterile. Greenblatt and Bock (1967) screened monoploids on the basis of stomata size and density, pollen size and pollen abortion However, due to strong influence of genotype these methods are not reproducible and reliable enough for ploidy estimation universally.
2. **Flow cytometry**. In plants usually ploidy estimation is carried out by chromosome counting on microscope slides prepared from actively growing root tips However, sometimes even chromosome counting is complicated due to very small size of chromosomes, insufficient spreading on slides and low frequency of dividing cells. Under such situations flow cytometry can be used for determination of the nuclear DNA content in plants (Bohanec, 2003). The nuclear DNA content of nuclei in certain phases of the cell cycle is related to the ploidy level. Hence flow cytometry could be applied for ploidy confirmation in plants. The methodology is based on the use of DNA specific fluorochromes analysis of the relative fluorescence intensity of the stained nuclei. Hence the relative DNA content or genome size level of generative ploidy, nuclear replication state and endopolyploidy of an unknown sample may be determined only after a comparison with nuclei of a reference standard whose genome size is known. Because the DNA content of the nuclei is related to the ploidy level, therefore, flow cytometric determination of DNA content may be used as an alternative to chromosome counting and other conventional methods for ploidy determination (Dolezel, 1991). To estimate nuclear DNA content, suspensions of nuclei/and /or permeabilized cells are stained with DNA specific fluorochrome and amount of light emitted by each nucleus is quantified and the result of the analysis is displayed in form of histogram of relative fluorescence intensity, representing relative DNA content in the samples. However, there are certain limitations of this method as preparations of suspensions of intact cells and nuclei suitable for flow cytometry are achieved with utmost precaution to avoid under estimation of DNA.
3. **Genetic markers**. In case haploids are to be utilized in a breeding program or some genetic study they should be detected at various plant stages among the F1 from interspecific crosses. The detection of haploids is based upon certain basic principle/s, for example, if a recessive homozygous (aa) female parent is crossed with a homozygous dominant male (AA), the F_1 thus produced would be (Aa). Nevertheless, if this F1 is recessive, it might be of

maternal origin haploid provided it is not the case of pseudo dominance or dominant gene expression has not been suppressed by any suppressor gene or it is not a maternal apomictic. For proper screening of haploids, markers for all the developmental plant stage, preferably seed or seedling stage, should be available. The most used marker in maize is based on anthocyanine pigmentation, which is generated by a dominant allele of the R-navajo system (R1-nj). Nanda and chase (1966) were the first to use aleurone color in selection of haploids. This allele promotes anthocyanine pigmentation in endosperm and in the embryo of the diploid seeds. The haploid seeds show pigmentation only in the endosperm, with the embryo remaining white. Nevertheless such markers may also have variable expressiveness and so haploidy needs to be confirmed by other methods also.

Use of haploids in crop improvement

1. **Production of homozygous lines** In any breeding program to improve the autogamous crops, the ultimate objective is to restore the homozygosity and stabilization of improved genotype in the end of the breeding program which usually takes approximately 8-10 years in the segregating populations. However, through haploid production, employing anther/pollen culture followed by doubling of chromosomes by colchicine, the homozygosity and stability can be restored within half of the time spent on traditional methods. Senadhira et. al. (2002) reported to have developed a saliniy tolerant dihaploid rice variety through use of haploids.
2. **Mutation expression in haploids**: Haploids can be utilized to observe the effect of induced mutations convenientlly since majority of the induced mutations are recessive in nature which are not expressed due to masking effect of dominant allele in diploids. To induce mutation pollen and single cells in a petri plates, like micro organisms, are exposed to mutation causing agents. Afterwards resultant mutant haploids are screened out for resistance to salts, herbicides, chilling, viruses, nematodes and nutritional deficiencies etc by providng such an environment/ s to develop haploids under investigation. The haploids exhibiting Improvement for specific traits/s are selected and diploidized by colchicine to be commercially used. InTobacco a mutant line resistant to black shank disease has been thus developed by employing such a methodology.
3. **Hybrid sorting:** Hybrid sorting for genetic improvement is carried out at gametic level since selected superior haploids are true representative of gametes. As a procedure, recurrent parent A and donor parent B are hybridized and their F_1 is raised on artificial media by rescuing the developing embryo. Anthers from this F_1 are used for anther culture to derive haploids

which represent a gametic array of F_1 hybrid and contribution of both A and B parental. Lines. The haplods having desired characters are subsequently selected and diploidized by colchicine application to make them homozygous, genetically stable and to be used as a commercial variety. This method is considered superior than pedigree and bulk method of selection since it provides opportunity to exercise selection at gametic level itself and consumes less time for inducing homozygosity (Fig 64).

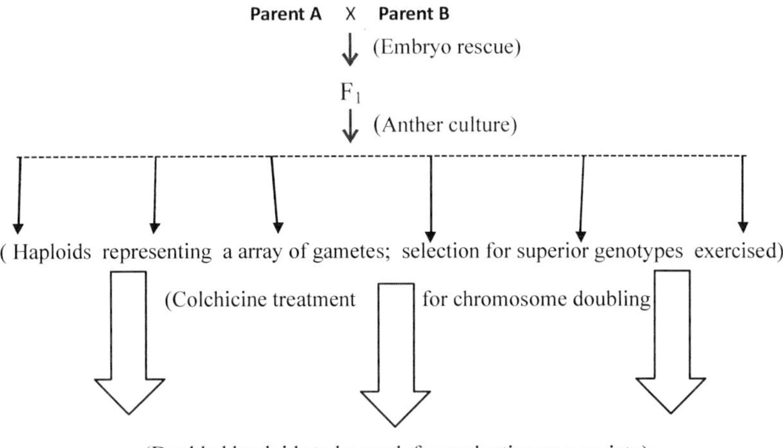

Fig. 64: Outlines of hybrid sorting methodology.

Thus developed haploids have been successfully used to develop varieties (double haploids) in various crops like barley, maize, sugarcane, oilseed, rape or canola *(B. napus)*, tobacco, rice and wheat.

5. Use of haploidy for genetic improvement of crops having polysomic inheritance

Improvement of the crops having polysomic inheritance is comparatively difficult since to select a desirable genotype from segregating population, a very large segregating population is required to be screened, to pickup preferred genotype/ s which is too cumbersome from practical point of view. Therefore, breeding at haploid level of such crops is more easy and convenient due to its simpler segregating ratios and small breeding population size, than at higher ploidy level. Chase (1963) suggested such a breeding scheme by scaling up and down of polyploidy referred to as **analytical breeding.** To elucidate the point an example of tetraploid potato (2n=4x=48) may be considered which has complex tetrasomic inheritance having probable genotypes as AAAa, AAaa or Aaaa. To improve traditionally such a crop is impossible since tetraploid potato genome contains high level of heterozygosity which may be lost upon sexual reproduction. However, analytical breeding based upon derivation of dihaploids from tetraploids facilitates the improvement to be carried out at derived diploids and consequently re-synthesizing tetraploids from these improved dihaploids overcoming the problems related to complex inheritance. To improve potato through analytical breeding following steps are taken:

1. First polyhaploids are derived from tetraploid cultivars The dihaploids derived from these genotypes may genotypically be AA, Aa and aa.
2. Derived dihaploids have simple disomic inheritance and hence they can be improved through selection growing smaller populations or can be crossed with their wild relatives to introgress new genes and variability.
3. Resynthesizing of tetraploids from these genetically improved dihaploids.

Aneuploidy. Trisomy, Monosomy and Nullisomy

Aneuploidy encompasses changes in the chromosome number less than the whole genome. It may result from subtraction (hypoploidy) or addition (hyperploidy) of a complete chromosome from the gain or loss of centromeres without significant change in the total amount of genetic material. The plants having polycentric chromosomes such as *Luzula* can increase the chromosome number by chromosomal fragmentation (agmatoploidy) without changing either the total chromosomal material or the overall number of centromere in the cell. The other important mechanism to create aneuploidy in diploids is **nondisjunction** during mitotic and meiotic cell divisions. Non disjunction during mitotic division leads to passing of sister chromatids to the same pole resulting in daughter cells with abnormal chromosome number (aneuplody). Similarly nondisjunction during meiosis leads to formation of unbalanced (n-1) and n+1) gametes rather than (n) which results into aneuploid progeny. The overall chromosome complement of an aneuploid is increased or decreased by one or more chromosomes by aneuploidy in plants, cells and tissues. The normal diploid chromosome number of an individual is denoted as 2n; so an aneuploid individual lacking one chromosome is denoted as (2n- 1) and called monosomic. Similarly the individuals deficient for a pair of homologues (2n-1") from their normal constitution are termed as nullisomic and the individuals having one additional chromosome or three representatives of a chromosome alongwith normal genetic constitution are (2n+ 1) are called trisomics. Individuals possessing two different extra chromosomes (2n+1+1) and having four representative of a chromosome (2n+2) are called double trisomic and tetrasomics, respectively. Contrarily to addition of chromosome/s their loss is usually more deleterious hence monosomics (2n-1) and nullisomics are rare in diploids but they occur in polyploids since the basic function of missing chromosomes can be carried out by remaining counterparts of additional genomes in polyploids. Due to such a compensating or buffering effect, monosomics and nullisomics series have been established in *Triticum aestivum, Nicotiana tabaccum, Avena sativa* and *Gossypium hirsutum. etc.*

(A) Trisomics

Blakeslee (1921a) first coined the term trisomic to denote the plants of *Datura* having three representatives of a particular chromosome (2n+1). The each chromosome of a haploid complement of Datura, in its turn could be involved in trisomic condition, hence primary trisomics of a particular species were equal to its haploid chromosome number, for example, in Datura n=12, hence only twelve different primary trisomics could be identified, one for each of the twelve chromosome types and each one of them had different size and shape of the seed capsule characteristically different in each of the twelve primary trisomics named as Rolled, Glossy, Buckling, Elongate, Echinus, Cocklebur, Monocarpic, Reduced, Poinsettia, Spinach, Globe and Ilex. Phynotypic differences in the trisomics also depend on ratio of the extra chromosome/s to the rest of the genome for example when chromosome 11 of Datura was added in one and two doses, the shape and size of resulting capsules were different from each other.

Origin of Trisomics

The trisomics may occur spontaneously in nature as observed in case of Datura. However, their occurrence is rare and so it was not possible to establish a full set of trisomics to be used for genetic studies. Hence a variety of artificial methods like use of triploids, x-rays and Gamma rays and desynaptic mutants were employed to produce different types of trisomics in different crops. Basically aneuploids originate due to following mechanisms.

1. **Non-disjunction or non-congression**. When chromosomes of a bivalent fail to separate at anaphase I, consequently the n-1 and n+1 gametes are produced which eventually give rrise a monosomic and and a trisomic after fusion with a normal n gamete. Belling and Blakeslee (1924) found cases of 11-13 disjuction and about 4% formation of n+1 pollen grains in Datura (2n=24). On the same pattern mitotic non- disjunction may lead to development of trisomic sector/s. during development.

2. **Asynapsis :** Some plants are homozygous recessive for certain of the factors which inhibit proper chromosome pairing or asynapsis and as a result a variable number of univalents are produced at metaphase I. These univalents segregate randomly to the poles leading to the formation of n+1 gametes. If these n+1 gametes are functional they may give rise trisomics in the progeny

after fusion with normal gamete. In a number of crops like maize and cotton, trisomics have been produced by exploiting asynapsis (Beadle, 1930) and (Beasley and Brown, 1942), In wheat, there are genes (5B complex) which inhibit the pairing between homoeologous chromosomes, if these genes are lost or suppressed, multivalent formation occurs leading to formation of some n+1 gametes and trisomics (Riley and chapman, 1958 and Sears 1976).

3. **Triploids**: Triploids are an important source to produce trisomics since during meiosis triploids form trivalents, bivalents and univalents resulting into abnormal meiosis and producing gametes having chromosome number varying from n to 2n. Out of these abnormal gametes some of them are obviously n+1 which may produce trisomics. Satina *et al.* (1938) found 48. 4% trisomics in the progeny of 3n x 2n cross of Datura (Table 12).

Table 12: Chromosome types in the progeny of 3n x 2n Datura)

Total	No. of plants	2n	2n+1	2n+1+1	2n+1+1+1
	285	58	*138*	79	10

4. **Translocation heterozygote:** The chromosomes of a translocation heterozygote during meiosis form a ring of four involving four chromosomes. These four chromosomes segregate in 3:1 rather than 2:2. The gametes which receive 3 chromosomes give rise trisomics particularly tertiary and interchange trisomics. To obtain primary trisomics interchange heterozygote should be crossed to the normal line. However, along with primary trisomics other types like tertiary and secondaries may also be produced hence separation of primary trisomics from others is made. The primaries thus produced contain normal counterpart of the interchanged chromosomes.

5. **Tetrasomics**. Tetrasomics are also good source for producing trisomics of a particular chromosome. Tetrasomic x diploid is reported to give higher percentage of primary trisomics in their progeny (Blakeslee and Avery, 1938)

Types of trisomics

1. **Primary trisomics (2n+1) :** The extra chromosome in these trisomics are normal one. Thus they contain completely homologous three chromosomes in their genetic constitution which can be denoted as:

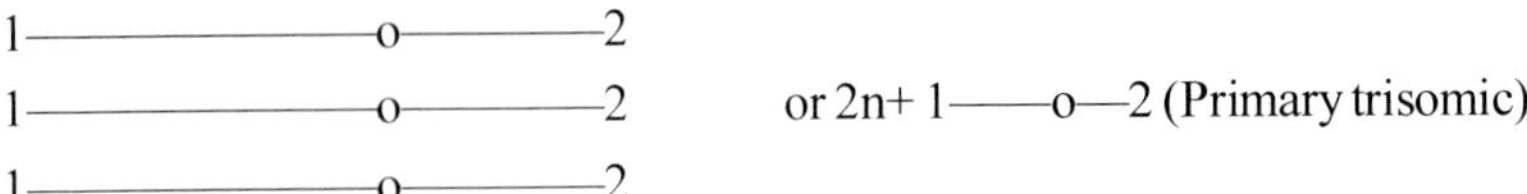

2. **Secondary and telocentric trisomics**: (2n+ iso chromosome and 2n+telocentric). The extra chromosome in these trisomics happen to be a iso-chromosome and a telocentric one, respectively. Iso-chromosome and

telocentric chromosome originate due to vertical mis-division of a chromosome generating telo-centric and iso chromosome (both arms genetically similar) (Fig. 65).

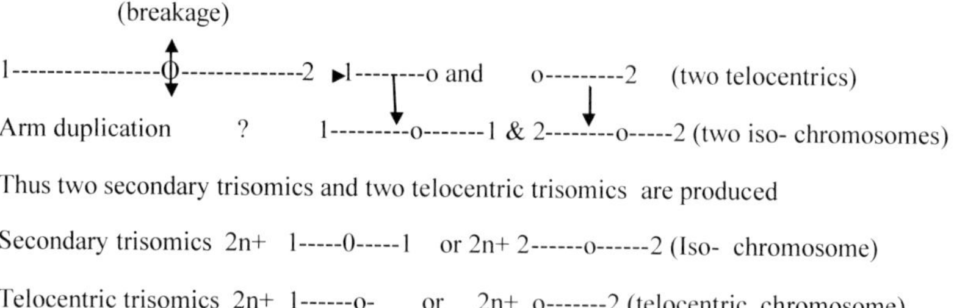

Fig. 65: Origin of secondary and telocentric trisomics, through mis-division and duplication of arm of chromosome

3. **Tertiary trsomics**:The extra chromosome in tertiary trisomics is translocated one and these trisomics originate basically among the progeny of interchange heterozygotes due to 3: 1 segregation of chromosomes to poles in a ring of four. Figure. 66 shows possible combinations of three chromosomes which may pass to one pole and one to the other The possibilities numbered I and II of chromosomal combinations when combine with normal gamete produce **interchange trisomics** and numbered III and IV produce plants with one translocated chromosome, known as **tertiary trisomics**. However, usually n+1 spores are prone to abortion and hence fail to be transmitted through male hence to produce tertiary triomics atleast they should be functional and be transmitted through female parent. Usually interchange heterozygotes with a ring of four showed 50% spore abortion, but in an interchange trisomic in corn 25 to30% spore abortion was observed (Burnham, 1930).

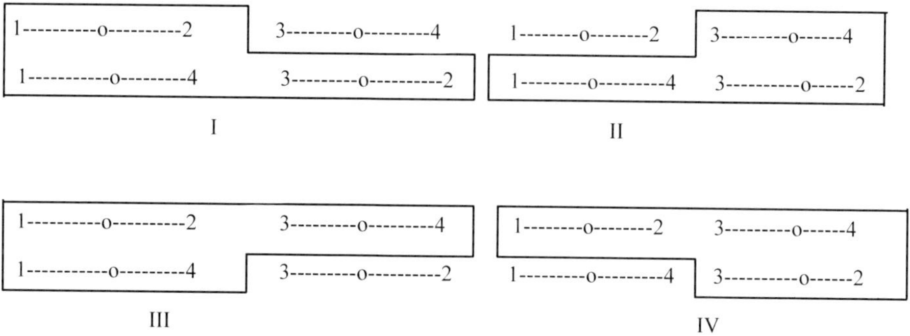

Fig. 66: Segregation of chromosomes (3:1) in a ring of four(⊖4) which produces tertiary trisomics

4. **Compensating trisomics**: In compensating trisomics one standard chromosome is missing which is compensated by the presence of other two chromosomes together. Such type of trisomics having 25 chromosomes (11^{II}+ 1. 2 +1. 9+ 2. 5) in *Datura (2n =24)* were observed in progeny of radium treated plants. In the given genetic constitution a standard chromosome 1. 2 is missing and it is being compensated by other two chromosomes, ie, 1. 9 and 2. 5. In other words chromosome 1. 2 is present in chromosomes 1. 9 and 2. 5 and thus the plant is only trisomic for chromosome segment. 9 and. 5. The compensating trisomics usually occur among the progeny of plants having a ring of (⊙6). The occurrence of ring of 6 takes place in the hybrid of two interchanges having a chromosome commonly involved in both translocations which consequently forms a ring of 6 during meiosis (Fig 67). When a gamete having specific set of chromosomes (10-9, 1-9, 5-6, 5-2) from ring of six combined with a normal gamete 10-9, 1-2 and 5-6) the compensatory trisomics were produced. A chain of seven chromosomes (9—o—10, 10—o—9, 9—o—1, 1—o—2, 2—o—5, 5—o—6, 6—o—5) is observed at metaphase I.

First interchasnge

I 10——o—-↓—9 II 1——o—-↓—2 III 5—o——6 Normal chromosomes

(↓) chromosomal break leading to translocation between I & II

(I) 10——————o———2 II 1———o———9 5———o————6

Second interchange between II and III chromosome

I 10——o——9 II 1——o—-↓——2 III 5——o—-↓——6 Normal chromosomes

(II) 10——o——9 II 1———o———6 III 5———O——2

These chromosomes of I and II interchange form a ring of 6 during meiosis (Fig. 67).

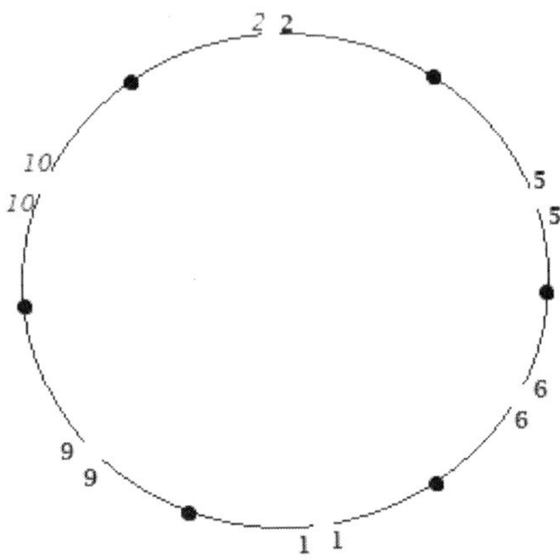

Fig. 67 : A ring of 6 in the F_1 hybrid of two interchanges during meiosis. When a gamete receives chromosomes 10. 9, 1. 9, 5. 6 and 5. 2 from this ring and combine with normal gamete having chromosomes 10. 9, 1. 2 and 5. 6, a tertiary trisomic (11II+1. 2+1. 9+2. 5) is produced where missing 1. 2 is compensated by 1. 9 and 2. 5

5. Telocentric trisomics: An individual with normal chromosome complement with an additional telocentric chromosome (2n+ 1telo (——o) are termed as telocentric trisomics. As stated earlier they originate due to mis- division of a chromosome and it is homologous to one arm of its standard counterpart. They frequently occur in progenies of autotriploid and other trisomics in 1—o—2 crops such as barley, maize datura, tomato and wheat.

Meiotic behavior of various trisomics : The various types of chromosomal configurations are observed during late diakinesis of trisomics having three homologous chromosomes to undergo pairing. These different configurations are basically caused by location and number of chiasmata present in the three homologues (Fig. 68). However, configurations made by primary, secondary and tertiary trisomics vary from each other.

Name of trisomic	Trivalents	Configurations	$II^s + I^s$
Primary	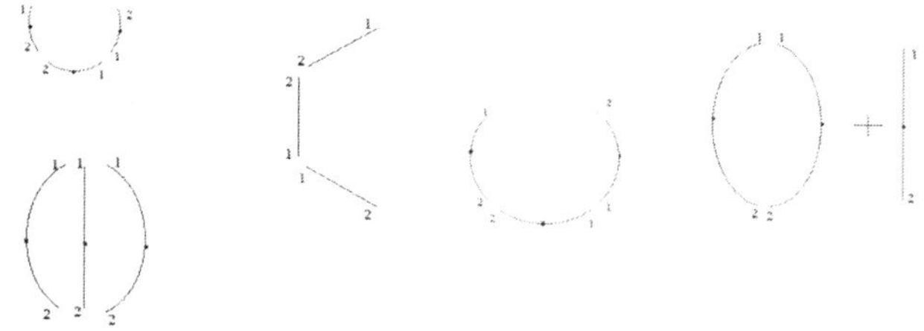		
1 ------- o------ 2			
1 ------- o------ 2			
1 ------- o------ 2			

Fig. 68 : Chromosome configurations in a primary trisomic at late diakinesis.

Genetics of Trisomics: In trisomics gene/s present on the extra chromosome are in triplicate hence they show polysomic inheritance in contrast to the diploids which show disomic inheritance. Trisomics produce n and n+1 types gametes rather only n gametes as in case of diploids. However, the transmission of n+1 gametes from male side is negligible and rarely exceeds 1% in most of the species but its transmission from female side occurs with varying degree, ie, 25% to 33 % of the n gamete, varying from chromosome to chromosome and with environmental conitionss. Khus (1973) observed in rice that gamete having extra chromosome (n+1) is not transmitted through male parent, but its transmission through female was good enough.

In diploids a pair of alternative alleles (Aa) produces two types of gametes, ie, A and a. But in case of trisomics (AAa or Aaa) four or five types of gametes (AA, Aa, aa, A and a) are produced. The gamete production has been described in detail in chapter autopolyploidy since gametes produced by a triploid and trisomic are identical for loci present on extra chromosome. The gametic out put is determined by presence or absence crossing over between gene and centromere to an large extent. If the gene is tightly linked to the centromere and crossing over is absent, the gametic output is determined by **chromosome segregation** (Table 4) and if the map distance between gene and centromere reaches 50 map units and croosing over is occurring with its all probability, it is determined by **maximum equational segregation** (Table 6) and if the degree of crossing over lies between these two extremes, the gametic out put is governed by **chromatid segregation** (Table 5). Furtthermore, multivalent formation (trivalents) and crossing over between gene and centrmere lead to formation of homozygous recessive gametes (aa) having identical alleles from same chromosome are produced by **double reduction** (Fig 53) Nevertheless, chromosomal segregation and maximum equational segregation are two hypothetical extremes to be fulfilled at biological level and hence actual experimental data falls between their expected ranges greatly depending upon the transmission of n+1 gamete/s, which varies organism to organism, environmental factors and with linkage intensity. If the gametic frequencies from **chromosome segregation, chromatid segregation and maximum equational segregation** are known, the phenotypic proportion of their progeny can be estimated in duplex (AAa) and simplex (Aaa) genotypes of trisomics. The nomenclature triplex (AAA), duplex (AAa), simplex (Aaa) and nulliplex (aaa) is based upon presence of number of dominant alleles in a particular genotype. The gametic output of various modes of segregation is listed in table 14, 15 and 16 be utilized to solve an actual experimental data of Mc Clintock and Hill, 1931, Table 13.

Table 13: Data from various tests of plants Trisomic for chromosome 10 (Maize) and heterozygous for R vs. r. (RRr) (From Mc clintock and Hill, 1931)

Cross/self	Colored aleurone	Colorless	Total	Observed % colorless	Observed ratio
Rrr					
1 RRr selfed	396	41	437	9.4	10:1
2 RRrx rr	819	213	1032	20.6	4:1
3 rr x RRr	941	486	1427	34.1	2:1
4 Rrr					
Rrr x rr	679	836	1515	55.2	1:1
Rr x Rrr	1392	2685	4077	65.9	1:2
Checks (2n)					
Rr selfed	608	204	612	25.1	3:1
Rr x rr	1161	1196	2357	50.7	1:1
rr x Rr	132	135	267	50.6	1:1

Types and frequency of gametes produced by chromosome, chromatid and maximum equational segregations: (A=R and a=r)

Table 14 : Expected gametes and their ratio resulting from chromosome segregation in a autotriploid /trisomic.

Transmission of n+1 gametes	Genotype	Gametes (n+1) (n)	Gametic ratio (AA:Aa:aa):(aa:+a)
50% (n+1)	AAa	1AA:2Aa ; 2A:1a	5:1
	Aaa	2Aa:1aa ; 1A:2a	1:1
25% (n+1)	AAa	1AA:2Aa ;3(2A:1a)=6A:3a	3:1
	Aaa	2Aa:1aa; 3(1A:2a)=3A:6a)	5:7

Table 15: Expected gametes and their ratio followed by chromatid segregation in a trisomic

Transmission of (n+1) gametes	Genotype	Gametes (n+1)	(n)	Gametic ratio (Aa;Aa:aa):(aa:a) (Dominant:recessive)
50%	AAa	6AA:8Aa:1aa	10A;5aa*	4:1
	Aaa	1AA:8Aa:1aa	5A:10a*	7:8
25%	AAa	6AA:8Aa:1aa	30A:15a#	11:4
	Aaa	1AA:8Aa:6aa	15A: 30a#	2:3

Adjusted according to n+1 gametes and their transmission frequency (#).

Table16: Expected gametes and their ratio produced by maximum equational segregation in an autotriploid/trisomic.

Transmission of (n+1) gametes	Genotypes	Gametes (n+1)	(n)	Gametic ratio (AA+Aa+A) : (aa+a)
50%	AAa	5AA:6Aa:1aa	8A:4a	19:5
	Aaa	1AA:6Aa:5aa	4A+8a	11:13
25%	AAa	5AA:6Aa:1aa;	3(8A:4a)=24A:12a	35:13
	Aaa	1AA:6Aa:5aa;	3(4A+8a)=12A:24a	19:29

Solutions of data obtained by Mc Clintock and Hill (1931) Table 13

1. RRr (selfed) {50% transmission, maximum equational segregation}

⚥	2R	1r
5RR	10RRR	5RRr
6Rr	12RRr	6Rrr
1rr	2Rrr	1rrr
24R	48RR	24Rr
12r	24 Rr	12rr

131R:13r = 10R:1r

2. RRr x rr {25% transmission, chromatid segregation}

⚥	r
6RR	6RRr
8Rr	8Rrr
1rr	1rrr
10R	10Rr
5r	5rr

24R:6r=4R:1r

3. rr x RRr

⚥	2R	1r
r	2Rr	1rr

2R:1r

4. Rrr x rr (50% chromosome segregation)

⚥	r
2Rr	2Rrr
1rr	1rrr
1R	1Rr
2r	2rr

3R : 3r= 1R:1r

5. rr x Rrr

⚥	1 R	2r
r	1 Rr	2 rr

1R: 2r

5. Checks (2n)

5A. Rr (Selfed)

⚥	R	r
R	RR	Rr
r	Rr	rr

3R;1r

5B. Rr x rr (BC)

⚥	r
R	Rr
r	rr

5C. rr x Rr (BC through male)

⚥	R	r
r	Rr	rr

1R :1r

The ratios for duplex (RRr) plants selfed for colored : colorless was 10:1 and for backcrosses the ratio was 4:1 through female, 2:1 through male, These ratios deviated significantly from the observed ratios in disomic plants, 3:1 in $F_{2,}$ and 1:1 in back crosses. The data from tests of simplex (Rrr) plants show about 1:1 ratio from Rrr x rr and 2:1from rr x RrrThe back cross data using trisomic as pollen parent and also selfed ones gives ratios that are markedly different from disomic plants. The trisomic ratio 0:1 for R gene establishes its location on chromosome 10.

Uses of trisomics

1. To localize a gene on a specific chromosome

Trisomics, specifically primary ones, have been extensively used to localize genes on specific chromosomes in several crops by several workers. The initial work in this field was carried out with *Datura stramonium* (Blakeslee, 1922, Blakeslee and Belling, 1924), Maize (Mc Clintock, 1929) and tomato (Leslley, 1932). Hermsen, (1970) provided basic information to localize monogenic, complementary, epistatic and, polymeric genes with the help of primary trisomics. However, in the present context the simple procedure to localize monogenic characters is being outlined based on the principle of polysomic segregation in **critical cross** due to presence of gene in question on a trisome. As discussed earlier the trisomic ratios result from three types of segregation, ie, chromosome segregation, chromatid segregation and maximum equational segregation, double reduction and different rates of female and male transmission of extra chromosome.

The basic requirement, to localize a gene on a specific chromosome of a species, is the availability of trisomic series for its constituent chromosomes which should be distinguishable from each other either morphologically or cytologically. As a general procedure trisomics, homozygous for the gene to be localized, are pollinated with a diploid. If the trisomic has dominant (RRR) genes, the diploid parent should be recessive (rr) whereas a dominant diploid dominant (RR) should be used as a pollen parent in case of recessive trisomics (rrr). The F_1 s thus produced is expected to consisists of disomic and trisomic plants. Theoritically the ratio of disomics and trisomics should be in 1:1 proportion since n and n+1 gametes are produced by female parent in equal proportion and are equally transmisable. The F_1 trisomics must be identified either morphololologically or by cytological checking to be selfed to produce an F_2 or to be backcrossed with a recessive diploid preferably from the stock from which the trisomics originated. If the character is based on one gene (monogenic), the normal ratios 3:1 and 1:1 will be observed in all F_2 and backcross populations, respectively except the line trisomic for the chromosome carrying the locus of the gene to be localized. In this case ratios will be modified since gene to be localized is present in triplicate. (polygenic inheritance). These critical ratios can be distinguished from normal ratios by application of chi-suare test.

To illucidate the procedure further we can consider an example where R locus present in a trisomic homozygous for dominant genes (RRR). To ascertain the location of R gene, it is (RRR) pollinated by recessive diploid (rr). In this case non critical trisomics will give rise normal ratios, ie, RR x rr →Rr ; Rr x rr→1Rr: 1rr (BC) and when selfed →1RR, 2Rr and 1 rr (3:1). In case of **critical cross** RRR x rr, two types of the F_1 plants, ie, disomic Rr and trisomics RRr are produced. Consequent upon back crossing or selfing of (RRr), the resultant ratios will deviate from **1:1 and 3:1**. Since gametic out put of RRr would be determined either of the three modes of segregation, ie, chromosome (1RR, 2Rr, 2R and 1r) chromatid (6RR, 8Rr, 1rr, 2R and1r) or maximum equational segregation 5RR:6Rr:1rr and 8R:4r) depending upon absence or presence of crossing over. and their transmission from female and male parent. The procedure of estimation of phenotypic ratios has been oulined to solve the data obtained by Mc Clintock in maize (Table 13). The deviation from normal backross ratio (I:1) and selfed ratio (3:1) indicates about the location of locus R on the extra chromosome (trisome) whose identity is already known. Chen and Sears (1967) localized gl gene of Arabidopsis on the chromosome 3 by use of trisomics. Their data presented in Fig. 69 clearly establishes the location of gl gene on chromosome3.

	Gl==O====		
(2n+1)	Gl==O====	X	gl----o---- (2n)
	Gl==O===	↓	gl----o----

Trisomics representing linkage groups

Total off spring combined	2 +: gl	3 +:gl	4 +:gl	5 +:gl
Proportion observed in F2	131:48 (2.7:1)	498:50 (10.0:1) *	49:15 3.3:1	94:27 (3.5:1)
Expected ratio	(3:1)	(17:1-8:1)	(3:1)	(3:1)

Linkage group 2, 4 and 5 show disomic raio and linkage group 3 shows modified ratio of 10:1, hence gl gene can be localized on chromosome 3.

* Deviated ratio from 3:1 and 1:1

: S. Lee-Chen and L. M. Steinitz-Sears (1967) Can. J. Genet. Cytol. 9: 381-384.

Based upon the same principle of distorted segregation gene/s can be localized on specific arm of a chromosome or on chromosomal segment by using secondary, tertiary, telo centric and acrocentric trisomics.

2. Hybrid seed production

To produce hybrid seed a breeder either opts the procedure of emasculation for making a plant male sterile or genetic methods for development of male sterile lines to be used as female parent for hybrid seed production Both the methods have their own limitations. Ramage (1965) suggested use of balanced tertiary trisomics having an extra translocated chromosome (tertiary trisomic) which continuously supplies male sterile lines for hybrid seed production in barley. A gene (ms) responsible for male sterility is located on normal chromosomes of BTT while its dominant counterpart Ms, closely linked to interchange break point, is located on translocated chromosome. Such a gene construct causes male sterility in all diploid progeny of BTT but BTT itself remains male fertile due presence of its dominant allele (Ms). During meiotic division BTT produces two types of gametes, ie, n and n+1 The tertiary chromosome included in n+1 gamete is only transmitted through female parent.

The tertiary chromosome also contains a dominant allele (R) of a marker gene either located very near to centromere or on break point of interchanged segment. Marker gene facilitates the identification of diploid plants in a population of diploids and balanced tertiary trisomics. Such marker/s could be red colored plant at seedling stage, plant height or a system of two genes controlling resistance and susceptibility. The basic principle to pr*oduce the hybrid seed by balanced tertiary trisomics lies in the breeding behavior of tertiary trisomics*. The trisomic produces n and n+1 gametes; both of which are equally transmitted by female parent but through male onle n gametes are transmitted producingt two types of progeny, ie, diploid (70%) and trisomics (30%). The presence of ms gene in diploids makes them male sterile while dominant Ms gene in trisomics confers male fertility (Fig. 68A).

(a) Breeding behavior of a Tertiary trisomic

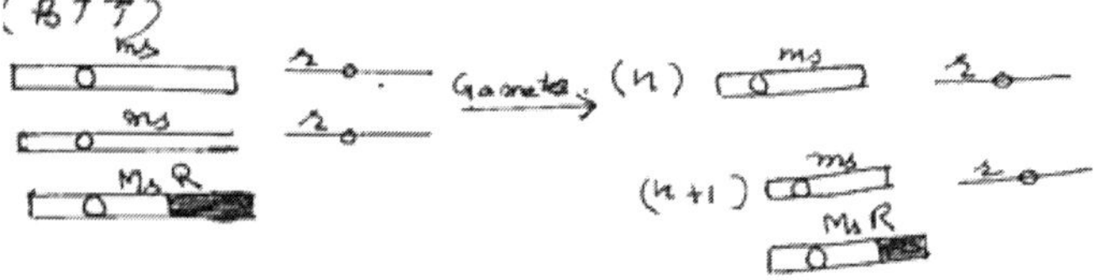

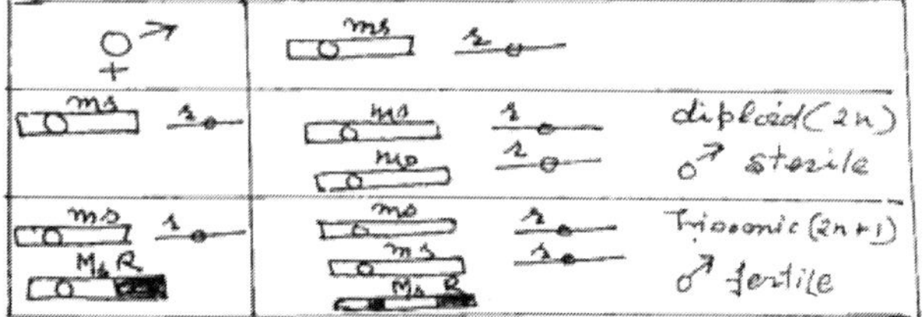

ms = male sterile Ms = Male fertile

Fig. 68A: Selfing of balanced tertiary trisomic produces fertile and sterile progeny in 1:1 ratio. The sterile progeny progeny serves the purpose of female parent for hybrid seed production

(b) The production of hybrid seed

1 BTT multiplication : To produce and increase the quantity of of BTT seed, the selfed seed of BTT is planted in separate field. It comprises of both diploids and trisomics (BTT). With the help of marker gene the diploids are identified and rogued out from the population at an early plant stage and only BTT seed is retained. Thus obtained some of the BTT seed is recycled for the same step and rest is used as a seed for female rows for hybrid seed production (Fig 68B).

Balenced Tertiary Trisomic Production

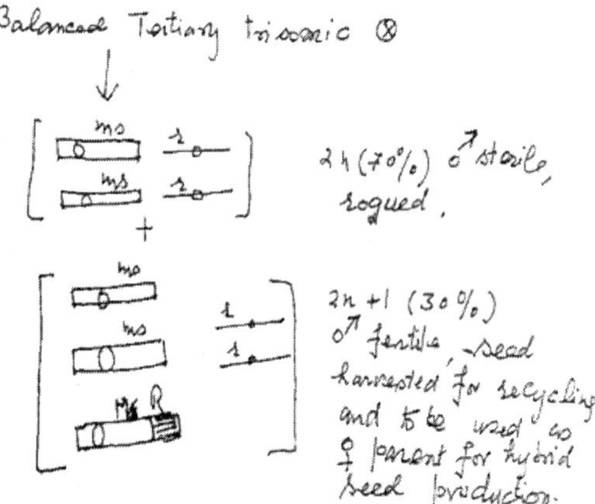

Fig. 68B: Multiplication and separation of BTT seed to grow as female lines

(c) Crossing block for hybrid seed production

In crossing block alternate rows of female parent, ie, seed from BTT and male rows (pollinator) are alternately planted. The female lines consists of fertile trisomics and male sterile diploids. The trisomics identified by markers are removed at an early stage and remaining male sterile diploids are used as female parent to be pollinated by pollen parent. (Fig. 68C).

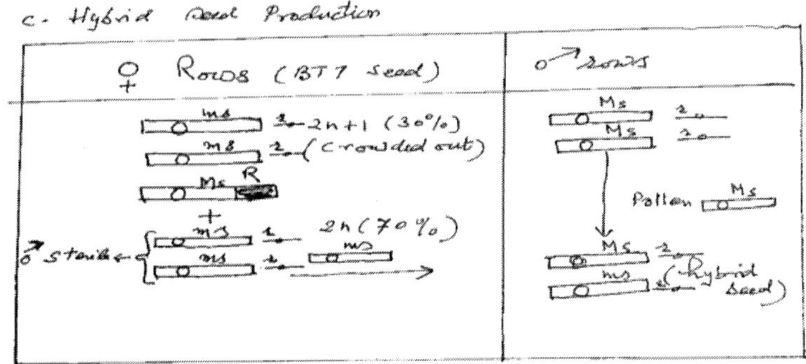

Fig. 68C: Method to produce hybrid seed in crossing block

(d) Alternative method to produce female seed

For easy and continuous supply of female seed, alternate rows of male sterile seed and BTT rows are planted. Since progeny of BTT consists of sterile diploid plants and fertile trisomics. The diploids hereby are rogued out at anearly stage and only BTT plants are kept to be used as male parent. Since only n gamete from BTT male parent is functional hence progeny thus produced is genetically male sterile which can be used further as a female parent (Fig. 68 D).

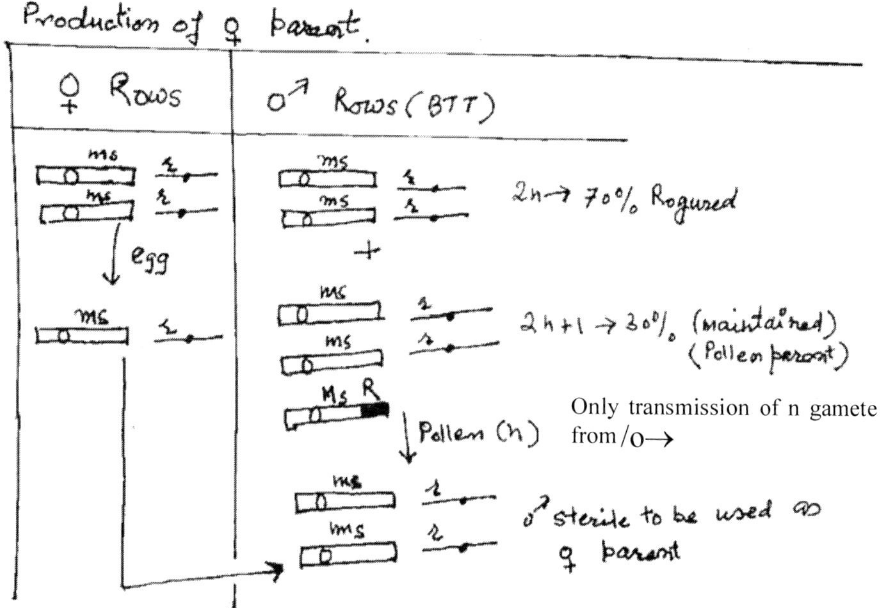

Fig. 68D: Easy and quick method to produce female parent

(B) Monosomy

Production of monosomics, assigning of a monosomic chromosome to its respective genome, breeding behavior of monosomics and their use in gene location

Monosomics: The individuals lacking one complete chromosome from their standard chromosome complement are called **monosomiocs (2n-1)**. If they are deficient of two different such chromosomes (2n-1-1), they are termed as **double monosomics**. The monosomics can be produced in polyploids rather than diploids since in polyploids missing function of deficient chromosome can be compensated by its counterpart present in additional genome. The monosomics were firstly reported in *Nicotiana tabaccum* by Clausen and Goodspeed (1926) and Later in 1944. Clausen and Cameron reported 24 monosomics in *Gossypium hirsutum*. To develop an insight about monosomics; tobacco and wheat monosomics will be dealt in detail in the present context.

Production of monosomics

1. **Use of haploids** Sears (1939) reported monosomic plants in progeny of haploid and normal wheat plants where as haploids were reported to occur in crosses of a wheat variety Chinese spring and rye *(Secale cereale* L.). The selfing of monosomics also led to production of other aneuploids like trisomics, nullisomics and tetrasomics. Matsumura (1940) established monosomics and nullisomics for the chromosomes of D genome of wheat by crossing species differing in chromosome number and by back crossing their F_1 to the parent having higher chromosome numbe. The methodology to produce monosomics by Sears (1939) is given below :

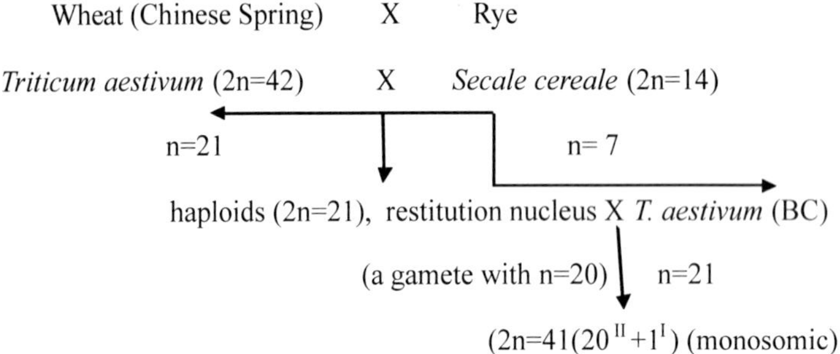

2. **From back crosses of interspecific hybrids:** Clausen and Cameron (1944) isolated 24 different *N. tabaccum* primary type monosomics in the progeny of back crosses of [(N. *tabaccum (n=24) x N. sylvestris (n=12)] x N. tabaccum n=24.*

(A)

N. *tabaccum(n=24 (TTSS)) x N.sylvestris(n=12)(SS)*

n= 12 T+12S ↓ n=12S

F_1 2n=36 (12 ^{II} SS + 12 ^{I} T) x *N. tabaccum* (2n=48)

n =23 (12S+11T) ↓ n=24 (12S +12T)

F_1 =2n=47 (12^{II} SS+ 11^{II} ^{TT}+1 $^{I\ T}$ monosomic of *N. tomentosiformis*

(B)

N. *tabaccum(n=24 (TTSS)) x N. tomentosiformis (n=12)(TT)*

n = 12 T+12S ↓ n=12T

F_1 2n=36 (12 ^{II} TT + 12 ^{I} S) x *N. tabaccum*2n=48)(BC)

n=23 (12T+11S) ↓ n=24 (12S +12T)

F_1 =2n=47 (12^{II} TT+ 11^{II} SS+1 ^{I} S monosomic of *N .sylvestris*

(C) (Matsumura 1940) D genome monosomics in wheat.

Triticum aestivum (2n=42) AABBDD X *T. dicoccum* (2n=28)AABB

(n=21)`(ABD) (n=14) (AB)

F 1 2n= 35(14 ^{II} AABB +7 ^{I}D) X *T. aestivum* (2n=42) (BC)

(n=14-21) (n=21)

or Preferable gamete n=20

2n=41 (20 ^{II} +1 ^{ID}) (monosomic)

3. **Use of Partially asynaptic and desynaptic plants:** During meiotic division partially asynatic plants do not show complete bivalent formation and univalents are distributed at random at anaphase I. which causes formation of deficient gametes having missed a chromosome. These deficient gametes ultimately lead to cause monosomic condition. Partially asynaptic plants were used to produce monosomics in wheat and tobacco. In wheat nullisomic-3 are also partially asynaptic and were used as another source of monosomics. Li et al. (1945) reported a recessive desynatic gene in wheat as a good source of monosomic production which can be introduced, through several back crosses into a variety to establish its recessive stock to produce monosomics.
4. **Occasional monosomy**: Worland and Law (1985) reported spontaneous occurrence of aneuploidy in wheat specially carrying Norin 10 dwafing genes. Cytological analysis confirmed these aneuploids to be predominantly monosomics. Riley and Kimber (1961) also reported the occurrence of 2-3% aneuploidy mainly monosomics in commercial wheat varieties which were caused by failure of chromosomal pairying during meiosis causing formation of aberrant gametes, ie, n-1 or n+1. Occasional occurrence of aneuploidy was also been observed in other seed crops also like tobacco, cotton and oats etc.
5. **Irradiation Treatment**: In crops like cotton and oats, irradiation treatment of inflorescencewas also used as a method to establish monosomics. It is presumed that irradion caused non disjunction of a normal binvalent leading to produce n-1 gametes.

Aneuploidy in *Nicotiana tabaccum*

A complete set of 24 primary types monosomics in *N. tabaccum* were reported by Clausen and Cameron (1944) which were produced by procedure (back crossing) as described above. To maintain these individual monosomic stock, the female monosomic is crossed to standard normal male (♀ monosomic $_{X}$ ♂ standard plant) so that monosomic chromosome always comes from standard male parent and uniform genetic background is maintained The impact of monosomy on the various plant characters is less pronounced than that of trisomy. However, monosomics in *N. tabaccum* differ from each other and diploids in respect of corolla, calyx, capsule size, devemental rate, leaf shape, intensity of chlorophyll and other chacters.

Meiotic behavior, spore abortion and breeding behavior of monosomics *(N. tabaccum)* monosomics

The monosomics of tobacco plants (2n=47) show various types of chromosomal associations as **(23^{II} +1^{I}** (univalents in addition to the bivalents), 22^{II} +1^{III} (association with non homologous) chromosome), 22 ^{II} +3 ^{I} or 21^{II}+5^{I} (non-conjunction in one more bivalent/s, etc. at diakinesis., The frequency of trivalent forrmation and non conjunction *i*n monosomics haplo D and haplo S were reported 25% and 20%, respectively. The trivalent formation was absent in other monosomics but non-conjunction was present in all varying from 5%to 25% or more. The chromosome disjunction was such that about 75% microspore carried 23 chromosomes (n-1) and 25% (n+1) 24 chromosomes. Univalents were observed to lag during meiosis and formation of micronuclei at quartet stage.

A normal plant showed 3% aborted pollens while in case of monosomics few showed 3% to 22 % aborted pollens and remainder 62% to 86%. The pollens that were smaller and shrunken and aborted probably had a missing chromosome (n-1). The ovule abortion varied from 16% to 87%. If different monosomics are crossed as female with a normal male plant, the variable % of monosomics are produced which ranges from 5 to 82 % and transmission of n-1 gamete from male parent produces some nullisomic progeny.

Identification of monosomics in relation to the basic genomes

Nicotiana tabaccum (2n=48 SSTT) an allotetraploid consists of two genomes S and T contributed by two diploid ancestral species, ie, *N. sylvestris* (2n=24) and *N. tomentosiformis* (2n=24), respectively (Clausen (1928). Since gametophytic chromosome number being n=24; it is expected to produce 24 monosomics equally belonging to both diploid genomes, ie, S and T. These monosomics can be asaigned to their respective genomes by making crosses of each monosomic separately with each of the two diploid progenitors and studying the meiosis of their progeny having 2n=35 chromosomes. It can easily be understood with the help of an example when it is not known whether the monosomics belong to sylvestris or tomentosiformis group.

N. tabaccum (monosomic unknown) X N. sylvestris

(2n=47) n=23 (12 +11) or n=24 (12+12) ↓ (2n=24) n= 12

F_1 2n=36 (rejected)

F_1 2n=35 ($12^{II}+11^{I}$ or $11^{II}+13^{I}$)

The chromosomal association of $12^{II}+11^{I}$ means that all sylvestris chromosomes are present and monosomic or missing chromosome belongs to *tomentosiformis,* contrarily the presence of second chromosomal association, ie, $11^{II}+13^{I}$ establishes that it belongs to *sylvestris.* The monosomic to be identified can also be crossed with N. tomentosiformis and the results obtained could be interpreted on the same pattern. The study of individual monosomics established that chromosomes of *N. tomentosiformis* ranged from small to large in size while that of N. sylvestris ranged from medium small to large.

Monosomics in wheat *(Triticum aestivum)* Modern wheat cultivars belong to two main species (1) hexaploid bread wheat *Triticum aestivum* (2n=42) (AABBDD) and (2) tetraploid hard or durum type wheat *T. turgidum* (2n=28) (AABB) used for macaroni and low rising bread. The bread wheat consists of three genomes, ie, A, B and D contributed by different diploid progenitors considered to have monophyletic origin. The gametophytic chromosome number in bread wheat is twenty one (n=21) hence wheat monosomic stock would be comprised of 21 different monosomics, designated by roman numerals, ie, I to XIV for monosomics of A and B genomes and XV to XXI for belonging to D genome chromosomes. The methology to assiagn a monosomic its respective genome is same as in case of *N. tabaccum.* Crosses of the different monosomics of *T. aestivuim* with T. durum (AABB) are made to identify those belonging to A or B genomes by counting the pairs and univalents in F_1 as:

(1) Assigning of monosomic to AB genomes or to D genome

♀ 2n-1 ($20^{II}+1^{I}$)(AABBDD) ? X ♂ T. durum (2n=28) (14^{II}) (AABB)

n=21 or n=20 ↓ n=14

F_1 = (1) 2n=35 (rejected)

(2) 2n=34 ($14^{II}+6^{I}$).

or (13 II + 8 I)

The inference from the chromosomal association of {2n=34 ($14^{II}+6^{I}$) }can be drawn that all chromosomes of A and B genome are present and the monosomic belongs to D genome and in case of second association the monosomic either belonged to A or to B genome.

(2) Assigning a monosomic to A or B genome

The monosomics thus assigned to A and B genomes together may further be distinguished to their respective A or B genomes by the counting number of bivalens and univalents in meiosis of F_1 hybrids between synthesized AADD amphiploid and monosomic stock isolated earlier for monosomics of A and B genomes. But meiosis of thus produced hybrid was disturbed to conclude the results hence an alternative technique was employed. The corresponding ditelocentrics (20 ^{II}+ 1^{II} telo) of the monosomics to be identified were derived for A and B genomes and crossed with AADD to produce F_1 hybrids with 2n= 34+telo. The meiotic study of such a cross is conclusive, even if the meiosis is disturbed, since hybrid could be cytologically analysed for formation of a heteromorphic pair or singleness of telosomic. The heteromorphic bivalent can only occur when telocentric pairs with A genome chromosome (since B genome is not there). On the other hand singleness of telocentric or unpaired telosomic establishes that monosomic beloged to B genome as given below:

♀ Ditelosomic for corresponding monosomic (20 II + 1^{II}telo) x ♂ Synthetic AADD (2n=28)

n=20+telo ↓ n=14

F_1 (1)= 13II + 1 ^{II} heteromorphic +7 ^{I} (belonged A genome)

(2) 14 ^{II} +6 ^{I}+ 1telo (free) (monosomic belonged to B genome)

Chapman and Riley (1966) reconfirmed the monosomics belonging to A genome by crossing them with *T. monoccocum* (2n=14). The following F_1 chromosome associations confirmed the genomic identity (A or B genome) of unknown monosomics.

♀Ditelosomics (20 ^{II}+ 1 ^{II} telo) X ♂ *T. monococcum* (2n=14)

n=20+ 1 telo ↓ n=7

F_1 (1)= 7 ^{II}+13 I+ 1 ^{I} telo→ monosomics belong to B genome

F_1 (2)=6 ^{II}+1 ^{II} (heteromorphic)+14 ^{I} → monosomics belong to A genome

Breeding behavior of a wheat monosomic

The breeding behaviour of a wheat monosomic has been reported by Sears (1944, 1952, 1954). Based on the progeny of monosomic 5A (9) pollinated by a normal pollen as well its selfing, it was established that deficient n-1 gamete functioned in 75% of eggs and 4% in pollen. The very low functionality of deficient n-1 gamete from male side was attributed to its inability to compete with the normal n gamete during fertilization and low viability. Thus a generalized breeding behaviour of a wheat monosomic was established with these values of functional n-1 and n gamete from female and male side. The 75% n-1 and 25% n and 4% n-1 and 96% n gametes are functional from female and male side, respectively The expected frequencies of different types of progenies, ie, normal (24%), monosomic (73%) and nullisomic (3%) having different chromosome types were produced by selfing of a monosomic (Table17). Tsunewaki, 1960) reported frequency of monosomics from 57. 3to 81. 9% with an average of 72 % in 2n-1 x 2n crosses.

Table 17 : Frequencies of the different chromosome types produced upon selfing of a monosomic (2n-1) with the (n-1) transmission rates 75% and 4% through female and male.

♀/♂	**n=96%**	**n-1=4%**
n=25%	2n (24%)	2n-1(1%)
n-1=75%	2n-1(72%)	2n-2(3%)

Progenies = Normal 24%, monosomics 73% and nullisomics3%

Use/s of monosomics:1Gene location on a specific chromosome (monogenic traits)

Genes can be assigned to chromosomes by the use of monosomics in allopolyploids. The procedure slightly varies depending on the condition whether the gene on the monosome is dominant, recessive or hemizygous ineffective.

(a) Locating dominant gene in a monosomic through F_1 analysis

A dominant gene present in a monosomic can be localized on a specific chromosome by observing phenotypic ratios in F_1's derived by crossing each monosomic line (A -) as female parent with another normal line carrying recessive allele (aa). Out of 21 such crosses, 20 crosses (monosomics for different other chromosomes) the F_1 progeny will be heterozygous for (Aa) expressing dominant trait uniformly without any segregation. These lines are called non critical ones. However, remaining one F_1 line, being deficient for the chromosome, probabily carrying that gene, will express both dominant and recessive phenotypes in 1:3 ratio (dominant:recessive). This cross is called a critical cross. Since identity of deficient chromosome involved in critical cross was already known hence a conclusion can be drawn that spefic gene is located on this particular deficient chromosome basically responsible for 1:3 aberrant segregation in F_1 as shown below:

(b) Critical cross or wth the right monosomic

♀Monosomic (with dominant marker) X ♂normal variety (with recessive marker)
(A-) ↓ (aa)
Gametes (-) (A) (a)

F_1 s (Based on male and female functional gametes)

♀/♂	(a) 100%
(-) 75%	(a-) 75%
(A) 25%	(Aa) 25%

(a-) 75% and (Aa) 25% or 3recessive : 1dominant

(c) Non-critical cross (gene is not on the monosome)

Monosomic (with dominant marker) (AA) X normal variety (aa)(with recessive marker)
(monososomic for chromosome not carrying the gene) ↓ (aa)
Gametes (A) ↓ (a)
F_1 Aa (all dominant)

3. Locating recessive gene in monosomic stock through F_2 analysis

In case monosomic stock carries recessive gene (a-), it is crossed with a normal line having contrasting dominant alleles (AA). However, the F_1 in critical as well as non critical cross will be uniformly dominant without any segregation to draw any conclusion as in the above example. Hence, F_1 (A-) is carried further to F_2 level with or without cytological screening. The cytologically screened (λ) (A-) selfed F_1 is carried further to F_2 analysis. It will give the phenotypic ratio of 97:3 (dominant recessive) (Fig a). If F_2 is raised from F_1 without any cytological screening the phenotypic ratio approximately will be 10:1

(a) critical cross
♀Monosomic (a-) X ♂ normal variety (AA)
Gametes; (a) and (-) ↓ (A)
F_1 → A - (75%) and Aa (25%) (based on deficient gametes transmission from male and female)
↓ selfing ↓ selfing
F_2 ↓ F_2 3:1(25%) → **18.75 :6.25** (A) dominant and recessive

♀/♂	A(96 %) -	-- (4%)
A (25%)	AA (24%)	A -- (1%)
---- (75%)	A—(72%)	-- -- (3%

97A : 3 – (with cytological screening of F1)
A 97: - 3→(75%) = 72. 75 : 2. 25 (B) dominant and recessive
(A) + (B) = A 72. 75+18. 75 (91. 50) :
(a) 6. 25+2. 25 (8. 50) = **A: a = 10:1** (without cytolgical screening of F_1)

Non critical cross (The deficient chromosome does not carry the concerned gene)

♀ Monosomic (aa) X ♂ normal variety (AA)

(a) ↓ (A)

F 1= Aa→ F 2→**3A:1a**

Maintenance of monosomics: The progeny of monosomics are usually monosomics for the same chromosome that was deficient in original monosomic stock. However, occasionally the progeny of monosomic may become monosomic for another chromosome. This phenomenon is called univalent shift (Person, 1956). During meiosis monosomics form bivalents plus a univalent at metaphase I as $\mathbf{20^{II}+1^{I}}$ in wheat and $\mathbf{23^{II}+1^{I}}$ in tobacco monosomics. But sometimes instead of a univalent a trivalent can also be formed due to association of monosome with non homologous bivalent or more than one univalents are formed due to lack of pairing in a bivalent.. These meiotic abnormalities coupled with occasional non –disjunction in a bivalent, lagging and misdivision of univalents at M I and M II and some other physiological factors lead to produce gametes which are deficient for the chromosome other than the original monosome and produce monosomics for an other chromosome (Fig. 69).

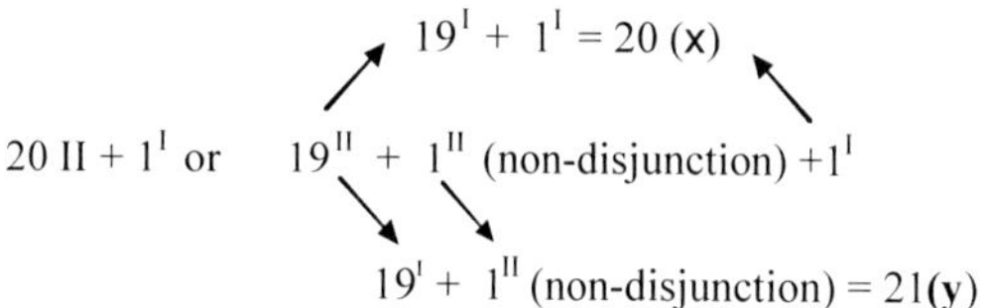

Fig. 69: Mechanism of univalent shift, the fusion of gamete (x), now deficient for the chromosome other than the original monosome. When it fuses with normal n gamete produces a monosomic for a different chromosome.

Therefore, utmost precaution must be taken in studies based upon monosomics. The identity of the monosome must be ascertained by proper chromosome counting and its morphology particularly its relative arm length, and total length. The identity of monosome can also be confirmed through its pairing with its counterpart telosomic chromosome.

Nullisomics

Nullisomics ($2n-1^{II}$) are those individuals which lack a pair of homologous chromosomes from their normal genetic constitution. Nullisomics are usually not found in progeny of diploid plants since missing vigour of deficient homologous pair can not be compensated in diploids. However, in allo-polyploids like wheat, having three genomes constituted of homoeologous chromosomes, nullisomic stock can be established easily due to their compensating capacity present among homoeologous groups. A complete set of nullisomics in bread wheat is available, however, they are not equally fertile and maintainable. About 50% nullisomics of wheat are either male or female sterile and the rest set some seed upon selfing and can be maintained as nullisomic lines. Only 8 wheat nullisomics were reported to be fairly stable (Sears, 1954) and even the progeny of thees had some aberrant plants.

Source of nullisomics

Nullisomics are obtained either from selfing of monosomics (Table 17) or from the progeny of monotelosomics and monoisosomics. Matsumura produced nullisomics for all the D genome chromosomes of wheat by using pentaploid wheat (AABBD) having disturbed meiosis. The seven D genome chromosomes in a pentaploid behave as univalents during meiosis which might produce a gamete having 20 (7A+7B+ 6 D) chromosomes. The union of such an egg and pollen having such a genetic constitution leads to produce a nullisomic individual.

Cytology of nullisomics

The nullisomics in wheat regularly produce 20 ^{IIs} and n-1 gametes during meiosis. However, nullisomics for majority of the chromosomes of the complement cause considerable meiotic irregularity, for an example, nullisomics for 2A and 3B of wheat are partially asynaptic (Sears, 1944). Ohta and Matsumura (1961) reported 20^{IIs} in about 90% of the microsporocytes of five nullisomics of D genome while in the remaining two, the frequency of cells with 20^{IIs} was lower. The other next frequent class other than 20^{IIs} happened to be 19^{II}+2 ^{Is}. One nullisomic of wheat, nulli- 5B, shows an effect opposite to that of asynapsis as absence of 5B in haploids greatly increased the frequency of bivalents and trivalents (Riley, 1958). As many as nineteen of the twenty chromosomes were commonly observed in various associations of 2, 4, 5, and 6 chromosomes.. More than one-half of the cells had at least one multivalent and many had several (Riley and Chapman, 1958 a).

Breeding behavior of nullisomics

About 50% nullisomics of wheat are either male or female sterile and the rest set some seed upon selfing and can be maintained as nullisomic lines. Only 8 wheat nullisomics were reported to be fairly stable (Sears, 1954) and even the progeny of these produce some aberrant plants having chromosomal association at metaphase I like 20 ^{II} ^{s}, 19^{IIs} + 1 ^{I} and 19^{IIs} +1 ^{III}. However, predominantly observed aberrant plants were nullisomic –trisomic (19^{IIs} +1^{III}). Usually chromosomes involved in nullisomic–trisomic condition were homoeologous which compensated for missing nullisomic chromosomes. Such plants are produced due to preferencial functioning of nulli-di (19+2) pollen produced due failure in pairying of a pair of homologous chromosomes and movement of these chromosomes to same pole. The wheat nullisomic 3B produce about 23.7% aberrant plants proving that all the gametes produced by a nullisomic are not always n-1 as expected from their normal meiosis. The crosses (nullisomuic x normal plant) are normally expected to produce only monosomic progeny but infact they also produce some other aberrant plants.

Morphological features of nullisomics: Nullisomics are identifiable from disomics on the basis of seedling vigour and other morphological features as reduction in floret number, degree of stiffness, degree of awning, compactness, size and appearance and, height and, tiller reduction.

Application of nullisomics

1. Chromosomal location of a gene

a) Absesence of dominant expression in the nullisomics: The applicability of this method is limited only to crop plants where monosomic series is available with dominant marker alongwith assured nullisomic survival. As earlier shown the selfing of monosomics produces disomics, monosomics and nullisonmics (24:73 :3). To locate a gene, all the monosomics are selfed and their progeny is observed. In the critical cross the nullisomics will express the absence of dominant expression of the particular gene/s. Since identity of the selfed monosomic was already known hence that gene can be assigned to that chromosome. Moreover, the dominant genes can also be as certained by their absence in particular nullisomics, for an example red seeded Chinese spring wheat has white seed in nulli -3D and absence of awns in nulli-4B and nulli-6B. Thus the gene for seed color and awn suppression can be assigned to chromosome 3D and 4B and 6B accordingly.

b) The location of a gene can also be determined by crossing a series of nullisomic lines with homozygous recessive for that gene. The majority crosses (non-critical ones) will exhibit dominant phenotype except one (critical One) which is deficient for that homologous pair on which the dominat allele is present.

1. Non-critical cross ♀ AA x ♂ aa

↓

F_1 Aa (all dominant)

2. Critical cross ♀ (- -) (nullisomic) x ♂ (aa)

↓

F_1 - a (all recessive) →

2. Establishment of homoeology

The nullisomic – tetrasomimic lines can be used to establish homoeology between genomes of wheat. The absence of a particular chromosome pair causes a change in normal phenotype while presence of an other extra or additional chromosome pair may compensate or restore the normal phenotype. For example, plants nullisomic for chromosome 20 in Chinese spring wheat are dwarf, having short and narrower leaves than normal and are male fertile and female sterile. But plants nullisomic for chromosome 20 and tetrasomic for chromosome 2 are normal and fertile. The same was true for plants nullisomic for 20 and tetrasomic for 13, leading to a conclusion that chrtomosome 2, 20 and 13 compensate among themselves while other chromosomes fail to do so. Therefore, these chromosomes are partially homologous or homoeologous. Based on the compensation tests homoeologus chromosomes of wheat are now numbered and assigned to A, B and D genomes of wheat (Table 9).

Variation in chromosome behavior (Endomitosis & Somatic segregation)

Endomitosis:Meiotic as well as mitotic cell division ensure chromosomal stability in living organisms in every generation. Since any deviation from the norm leads to undesirable genetic consequences. Nevertheless, there are certain mechanisms where such a rule is violated for the certain genetic function in plants as well as animals. Endo polyploidy and somatic segregation are the examples.

Endomitosis, as described by Geitler (1939), is one of the mechanism through which the cell systematically increases its chromosome constitution. It occurs as a normal feature in some tissues of higher plants resulting into mosaic of cells having different chromosomal complement (diploid and polyploid) for that species. Two fundamentally different types of endomitotic cycles exist which morphologically appear to be resembling with each other. In the first type as in the tapetal cells of the Liliaccous plant *Eremurus* and in the embryo suspensor cells of *Phaseolus,* the chromosomes decondense and recondense during the replication cycle and increases in number under endomitotic mechanism in the same cell (Brady and Clutter (1974). In the second type as in the septal cells of grasshopper testicular follicles, the nuclei which morphologically represent different endomitotic stages apparently are stationary and probably play the role of transcription of one or more gene product. The phenomenon of endomitosis is reported to be characteristic of non –meristematic and non-germline cells such as cortex, pith, xylem vessels and nutritive tissues like tapetum and endosperm

The discovery of polytene chromosomes of salivary glands of drosophila in 1930's happened to be pioneer research work in this field. In fruit fly (Drosophila)

mitotic divisions stops after 18 hours of larval development, but chromosomal DNA and cell growth continue apace in geometrical progression as 2→4→8→16 doubling series and thus as much as 1024 times haploid amount of DNA is synthesized in several copies of each chromosomes and all copies of the chromosome type remain together of all the 8 chromosomes in the nucleus. All chromosomes are thus amplified in size and all the homologous pair tend to be associated with one another. The phenomenon of such a pairing i s known as **somatic pairing** due to which the paired chromosomes look like four giant chromosomes. Thus selected cells of the Drosophila larvae exhibit **endopolyploidization** of the diploid genome leading to formation of polytene chromosomes. These endopolyploid cells never again divide and do not parcipate in pupation period of fruit fly but remain active in the sense that they respond to environmental stimuli and synthesize specific type of proteins.

The most extensive studies in field of endopolyploiudy have been made in insects and plants and a variety of endopolyploid cells have been described. Until recently the focus had been on morphological features and cytological abnormalties of the end. Now these cytological abnormalties are being associated with function (Nagl 1978, 1982) and amplified level of polyploidy in insects is reported to be responsible for differentiation of tissue. However, in case of plants the more common situation is that a tissue contains cells with different chromosome constitution, ie, a mosaic of diploid and polyploid cells. Now it is being argued that the same number of gene copies are achieved through various polyploidization mechanisms as can be through normal mitosis and cytokinesis However, the plus point of endopolyploidy to the organism is less time and energy required and that gene transcription can continue uninterrupted by mitosis and cell division (Nagl 1978). This interpretation agrees with the observations that high levels of polyploidy are often characteristic of gland and other cells which show intensive synthetic activity (Nagl 1978). A further phenomenon which agrees with these ideas concerning the significance of endopolyploidy is wide spread amplification or under replication of certain genes or chromosome segment/s. If only those genes whose transcription is important for the specific cell type occur in multiple copies or conversely are under replicated where they are not needed, this too would serve the economy of tissue and cell differentiation.

Mechanisms and types of endocycles

1. **Endopolyploidy:** The endopolyploidy results from the processes in which successive replication cycles occur (2 or more times) to replicate the chromosomes in absence of spindle. The most common of these is the endo- duplication in which chromosome replicate 2 or more times without intervening mitosis. This endo-duplication gives rise to polytene chromosomes

in which the chromatid remain attached to the ribbon like chromosomes having up to a thousand or more strands. The well known examples of endopolyploidy are salivary gland polytene chromosomes in Diptera insects and in many plants they are characteristic of endosperm, antipodal, haustrial and other cells of ovules.

2. **Endomitosis**: The other type of endocycle is **endomitosis** whose implies that chromosomes decondense and recondense during replication as in normal mitosis and these processes occur inside the nuclear membrane without formation of spindle fiber consequently leading to increase the number of chromosomes. Out of the four stages of endomitosis; its prophase is almost similar to that of normal mitotic prophase. During this period chromosome continue to condense and remain attached to nuclear membrane. In endo - metaphase the sister chromatids lie side by side. and subsequently their distance increases with the initiation of endo-anaphase and during endo-telophase they decondense to form interphase nucleus (Therman *et al.*, 1983). The common situation created by endo mitosis happens to be the presence of tissues having cells with different chromosome constitution, ie, mosaic of diploid and polyploid cells.

Somatic segregation

The somatic segregation has been defined as "the production of two genetically dissimilar cells at a somatic cell division" and as a result plants and animals exhibit different genotypic characters in various parts of the same individual. The examples of somatic segregation observed were half and half individuals, ie, gynandromorphs in insects, birds with varied plumage and fruits and flowers with stripes of different colors in plants. The vegetatively propogated plants are supposed to be genetically uniform, when grown in large numbers, have given many examples of mosaic pattern. Some of the changes in vegetatively propagated plants even have been selected to give rise new varieties. Since such deviation from normal generally occur after fertilization hence term "**somatic segregation**" is generally applied. However, no regular process is known to induce such genetic changes but variation thus observed is considered a deviation from normal. These changes are reported to occur at all stages of the development with higher frequency than earlier thought. The crossing over normally occur during meiosis and create variation, which also do occurs in somatic cells of some organisms More over exchange of chromosomal segments between non homologous chromosomes also do occur before meiosis generating visible effects after germ cell recombination. Along with somatic crossing over and reciprocal translocations, all the changes as chromosome loss or rearrangement capable of propogation are included as a cause/s of its origin. Nevertheless, little critical evidence of its origin has been given and even Babcock and Lloyd (1917) has argued that the use of term somatic segregation a misleading one used for variation in horticultural crop only. However, Burnham (1962) described an example of somatic segregation in a 42 chromosome stock of *Avena sativa* reported to segregate in normal and 40 chromosome dwarf. in 9:1 ratio. The normal 42 chromosome progeny futher segregated in the same ratio plants Out crossed seed produced only normal progeny but part of it again segregated dwarfs. Meiosis was normal in the plants whose progeny segregated, but root tip analysis of germinating cells revealed sterile dwarfs to have 40 chromosome having lost a particular recessive genotype particular chromosome pair between fertilization and maturation of the embryo, The basic cause of such a phenomenon was attributed to proneness of last mitotic division to non disjunction.

References

1. Avivi, A.G., Feldman, M. And Brown, M. (1982). Chromosoma 86: 1-26.
2. Babcock, E.B. and Lloyd, F.E. (1917). J. Hered. 8: 82-89.
3. Bauman, L.F. (1961). Agronomy Abstr.; 53rd Annual Meetings Amer. Soc. Agron.: 47.
4. Beadle, G.W. (1930). Carnell Univ. Agr. Exp. Stn. Mem. 129.
5. Beasley, J.O. and Brown, M.S. (1942). J. Agr. Res. 65: 421-427.
6. Belling, J. and Blakeslee, A.F. (1924). Proc. Natl. Acad. Sci. U.S.A. 10: 116-120.
7. Bennett, M.D.; Finch, R.A. and Barclay, I.R. (1976). Chromosoma 54: 175-200.
8. Blakeslee, A.F. and Belling, J. (1924). J. Hered. 15: 195-206.
9. Blakeslee, A.F. and Avery, A.G. (1937). Journal of Heredity. 28(12-1): 393-411.
10. Blakeslee, A.F. and Avery, A.G. (1938). Coop. in Res. CarnegieInst. Wash Pub. 501: 315-351.
11. Blakeslee, A.F. (1921a). Genetics 6: 241-264.
12. Bohanec, B. (2003). A Manual, Maluszynsk, M., Kasha, K.J., Forster, B.P. and Szarejko, I. pp. 397-403, Lluwer Academic Publishers, ISBN 1-4020-1544-5, Dordrecht.
13. Brady, T. and Clutter, M.E. (1974). Chromosoma 45: 63-79.
14. Brouwer, D.J. and Osborn, T.C. (1999). Theor. Appl. Genet. 21: 207-286.
15. Burnham, C.R. (1934). Genetics 19: 430-447.
16. Burnham, C.R. (1956). Bot. REV. 22: 419-552.
17. Burnham, C.R. (1962). Discussions in Cytogenetics. Burgess Publishing Company, Minneapolis.
18. Clausen, R.E. and Cameron, D.R. (1944). Genetics 29: 447-477.
19. Clausen, R.E. and Goodspeed, T.H. (1926). Univ. Calif. Pub. Bot. 11: 61-82.

20. Cleland, R.E. (1960). Z. Verebungslehre 91: 303-311.
21. Darlington, C.D. and La Caur, L.F. (1950). Heredity 4: 217-248.
22. Dodsworth, S.; Chase, M.W. and Leitch, A.R. (2016). Bot. J. Linn. Soc. 2016;180:1-5.
23. Dolezel, J. (1991). Phytochem Analysis (2): 143-154.
24. Doyle, C.G (1979). Theor. Appl. Genet. 54: 103-112.
25. Driscoll, C.J. and Jensen, N.F. (1964). Crop Sci. 4: 372-374.
26. Federrow, A. (1969). Chromosome Numbers of Flowering Plants, Komorov Bot. Inst; Leningrad.
27. Geitler, L. (1939). Chromosoma 1: 1-22.
28. Greenblatt, I.M. and Bock, M. (1967). J. Hered. 58: 9-13.
29. Guedes-Pinto, H; Darvey, N. and Carnde, V. (Eds.) (1996). Triticale Today and Tomorrow, Kluwer Academic Publishers.
30. Gustafson, J.P. and Zillinsky, (1973). Proc. 4th. nt. Wheat Genet. Symp.
31. Hagberg, A. (1962). Hereditas 48 (1-2) : 243-246.
32. Hermsen, J.G (1970). Euphytica 19: 125-140.
33. John, B. and Lewis, K.R. (1965). Protoplasmatalogia, VI, F1: 102-103.
34. Karpechenko, G.D. (1928). Zeit. Ind. Abst. Vererbungsl. 39: 1-7.
35. Kasha, K.J. (1974). In: Kasha, K.J. (ed.) Haploids in higher plants-Advances and Potential. Univ. of Guelph, Guelph (Canada) pp. 67-87.
36. Katiyar, R.K.;Chamolr, R. and Chopra, V.L. (1998). Plant Breeding 117: 398-399.
37. Kato, M. and Tokomasu, S. (1976). Euphytica 25: 761-767.
38. Khara, H. and Tsunewaki, K. (1962). Jap. J. Genet. 37: 313-313.
39. Khus, G.S., Rick, C.M. (1968b). Cytologia 33: 137-148.
40. Khus, G.S., Rick, C.M. (1967b). Genetics 38: 74-94.
41. Khus, G.S., Rick, C.M. and Robinson, R.W. (1964). Science 145: 1432-1434.
42. KIhara, H. (1944) Agr. and Hort. Japan 19: 889-900.
43. Kimber, G. and Sears, E.R. (1987). In: Heyne, E.G. (ed.) Wheat and Wheat Improvement, 2nd Edn. Am. Soc. Agron. Monogr. 13: 154-164.
44. Kimber, G. and Ril; ey, R. (1963). Bot. Rev. 29: 480-531.
45. Kimber, G. and Riley, R. (1963). Bot. Rev. 29: 480-531.
46. Knott, D.R. (1971). In: Mutation Breeding for Disease Resistance. I.A.E.A. Vienna. pp. 67-77.
47. Kush, G.S. (1973). Cytogenetics of Aneuploids, Academic Press, New York.

48. Laibach, F. (1925). Z. Bot. 17: 417-459.
49. Lande, R. (1983). Evolution 38(4): 743-752.
50. Larter, E.N. and Hsam, S.L.K. (1973). Canadian Journal of Genetics and Cytology. 15(1) : 197-204.
51. Lesley, J.W. (1932). Genetics17: 545-559.
52. Li, H.W. Pao, W.K. ad Li, C.H. (1945). Am. J. Bot. 32: 92-101.
53. Matsumura, S. (1940). Bot. Mag. Tokyo 54: 404-413.
54. Mc Clintock, B. (1929). J. Hered. 20: 12-18.
55. Mc Clintock, B. (1931). Missouri Agr. EXP. Sta. Bull. 163: 1-30.
56. Mc Clintock, B. (1933). Zeits. Zellforsch. u. mikr. Anat. 19: 199-237.
57. Mc Clintock, B. (1934). Zeits. Zellforsch. u. mikr. Anat. 21: 293-328.
58. Mc Clintock, B. and Hill, H. E. (1931). Genetics 16: 175-190.
59. Mehetre, S.S. and Thombre, M.V. (1981). Proc. indian Acad. Sci. (Plant Sci.). 90(4): 313-322.
60. Moore, R J. and Cave, M.S. (eds.) (1965-74). Index to chromosome numbers, Reg. Veg; Utrecht.
61. Nagl, W. (1978). Int. Rev. Cytol. 73: 21-53.
62. Nanda, D.K. and Chase, S.S. (1966). Crop Sci. Vol. 6: 213-215.
63. Nordenskjold, H. (1945). Acta Agri. Succanae 1: 1-136.
64. Ohta, T. and Matsumura, S. (1961). Cytologia 26: 226-235.
65. Person, C. (1956). Canad. J. Bot. 34: 60-70.
66. Ramage, R.T. (1965). Crop Sci. 5: 177-178.
67. Rick, C.M. and Khus, G. S. (1966). Heredity (Suppl.) 20: 8-20.
68. Riley, R. (1966a). Contemp. Agric. 11-12: 107-117.
69. Riley, R. (1966b). Proc. 2nd Int. Wheat Genet. Symp. Hereditas (Suppl.) 2: 395-408.
70. Riley, R. and Chapman, V. (1958). Nature, London, 182: 713-715.
71. Riley, R. and Chapman, (1958a). Heredity 12: 89-100.
72. Riley, R. (1958). Proc. X Int. Cong. Genetics, Montreal 2: 234-235.
73. Riley, R. and Chapman, V. (1958). Nature (Lond.) 182: 713-715.
74. Riley, R. and Kimber, G. (1961). Heredity 16: 275-290.
75. Sarashima, M. (1973). Special Bull. Coll. Agric. Utsunomiya Univ. 14: 11-26.
76. Satina, S; Blakeslee, A.F. and Avery, A.G. (1938). Am. J. Bot. 25: 595-602.
77. Sears, E.R. (1944). Genetics 29: 232-246.
78. Sears, E.R. (1952). Chromosoma 551-562.

79. Sears, E.R. (1954). Missouri Agr. Exp. Stn. Res. Bull. 572.

80. Sears, E.R. (1956). Brookhavren Symposia in Biology. Genetics in Plant Breeding: 1-22.

81. Sears, E. R. (1939). Genetics 24: 509-523.

82. Sears, E. R. (1956). Brookhaven Sym. Biol. 9: 1-22.

83. Sears, E. R. (1976). Ann. Rev. Genet. 10: 31-51.

84. Senadheera, D; Zapatha-Aris, F. J.; Gregorio, G. B.; Alejar, M. S.; Cruz, H. C; Padolina, T. F. and Galvez, A. M. (2002). Field Crops Res. 76: 103-110.

85. Sharma, D. and and Knott, D. R. (1966). Can. J. Genet. Cytol. 8: 137-143.

86. Singleton, W. R. (1939). Genetics 24: 109.

87. Stadler, L. J (1935). Amer. Nat. 69(80-81). (Abstr.)

88. Stadler, L. J. (1933). Missouri Agr. EXP. Sta. Bull. 204: 3-29.

89. Sturtevant, A. H. (1921). Proc. Natl. Acad. Sci. U. S. A. 7: 235-237.

90. Therman, E; Sarto, G. E. and Stubblefield, P. A. (1983). Human Genetics 63(1): 13-18.

91. Thomas, J. B. and Kaltsikes, P. J. (1972). Canadian Journal of Genetics and Cytology. 14 : 4889-898.

92. Tsunewaki, K. (1960). Jap. J. Genet; Japan 11: 46-47.

93. Wehner, T. C. (2008). Proc. Cucurbitaceae 2008 EUCARPIA meeting p. 79-89 (ed. M. Pitrat).

94. Wehner, T. C. (2008). Watermelon (381-418). In J. Prohens and F. Nuez (eds.) Handbook of Plant Breeding Vegetables I:Asterraceae, Brassicaceae, Chenopopodiaceae and Cucurbitaceae, Springer Science+Busness LLC, New York, NY, 426 P. 17.

95. White, G. B. (1978). Mosquito Systematics 10 (1) :13-31.

96. Worland, A. J. and Law, C. N. (1985). Euphytica 34: 317-327.

97. Zou, J[1]; Zhu, J; Huang, S; Tian, E; Xiao, Y; Fu, D; Tu, J; Fu, T. and Meng, J. (2010). Theor. Appl. Genet. 2010 Jan; 120(2) : 283-290.

4

Intervarietal Chromosome Substitution Lines

The intervarietal chromosome substitution lines of wheat *(T. aestivum)* are utilized to study the effects of individual whole chromosome/s on agronomic characters of interest to improve crop plants to be selected for preferred traits (Law and Worland 1973). The common method to develop subsitutiion lines involves development of monosomics lines first of that variety which are used later as recipient variety or recurrent parent for substitution. Presently monosomic lines of several wheat varieties are available. However, the development of substitution lines for a specific chromosome by using monosomics sometimes may be erroroneus due replacement of specific monosome by another one of the 21 chromosome set from the donor to the recipient and consequently monosomic becomes monosomic for another chromosome or a recombined chromosome (univalent shift.). Therefore, to maintain and confirm the identity of a chromosome involved in monosomy are tagged with either genetic marker/s or confirmed cytologically or with the help of molecular markers.

1. Transfer of monosomic condition from one variety to recipient variety

The procedure to transfer of monosomic condition from one variety to other was established by Sears (1953). To transfer monosomic condition from one variety to other, firstly these are hybridized and then repeated back crossing is used to restore the original genotype. The example below elucidates the transfer of monosomic condition from **Chinese spring** to **Kenya Farmer**.

♀Monosomic (Chinese spring) x ♂ Kenya Farmer

($20^{II\,ch} + 1^{Ich}$) $21^{II\,KF}$

Gametes→ n=21^{Ch}, n-1= 20^{ch} ↓ n= 21^{KF}

F $_1$ → (select monosomic) ($20^{ch} + 1^{KF} / 20^{KF}$) x$21^{II\,KF}$(repeated BC to reduce Chinese spring chromosomes)

(20 +1^{KF} / 20)→

→ The identity of the monosomic, ie, 1^{KF} is guarded alongwith repeated back crossing to restore the original genotype of Kenya Farmer having monosomic condition, ie,($20^{IIKF}+1^{KF}$). The procedure is followed to develop the set of 21 monosomics.There after these monosomics are used to develop complete series of substitution lines of Chinese spring having substituted a pair of chromosomes by its homologous counter part from Kenya Farmer ($20^{IIch}+1^{IIKF}$) is developed.

2. Development of substitution lines from monosomics

To develop the substitution line between a variety (a), a recipient monosomic, and variety (b) having contrasting characters, a disomic and donor, are crossed and their monosomic F_1 is selected to be back crossed to Chinese monosomic stock(a). To maintain the identity of univalent chromosome from variety (b), in subsequent back crosses, the selected monosomic is selfed to get disomic after each generation of back crossing to produce a substitution line. The disomic, thus produced, is always is used as a male parent with the same monosomic.

(a) ♀ Recurrent variety (a) (monosomic) (2n-1) x ♂ non recurrent variety (b) (2n)

(n & n-1) ↓ (n)

(I) F1 s → 2n (discarded) and 2n-1 ($20^{II\,ab}+^{1\,b}$) →selected & selfed→ ($20^{IIab}+1^{IIbb}$ disomic)

(II) (BC to (a) monosomic x $20^{IIab}+1^{IIbb}$ (disomic)

↓ (after repeatedl backcrosses between monosomic and disomic)

F1= $20^{II\,a}+1^{Ib}$ → selfing → $20^{II\,a}+1^{II\,b}$ (a substitution line)

The selected monosomic will have replaced the chromosome of (b)out of 20 bivalents but always having monosomic chromosome from variety (b) and ultimately after repeated back crosses (6 to 8), the constitution of monosomic will be ($20^{II\,aa}+1^{Ib}$) which after selfing would be a substitution line having constitution $20^{II\,aa}+1^{IIbb}$, a substitution line

3. Development of substitution lines by nullisomics

Nulli-Chinese ($20\ II^{Ch}$) (CS) x Kenya Farmer ($21\ II^{KF}$)→F_1 (20^{CH}// $20^{KF}+1^{KF}$) (monosomic)

Backcross nulli–chinese ($20\ II^{Ch}$) x (20^{CH}// $20^{KF}+1^{KF}$) (monosomic) → each back cross will reduce the number of KF chromosomes while keeping univalent 1^{KF} intact. Eventually the original chromosomal constitution of monosomic (99.6%) will be ($20^{IIch}+1^{KF}$). Thus obtained monosomic upon selfing produces a substitution line having chromosomal composition ($20\ II^{ch}+1^{IIKF}$). The details of the procedure has been outlined below.

(1) ♀ Chinese (20 II^{CH}) X♂ Kenya Farmer (21 II ^{KF})

↓

F_1 (20 ^{CH}// 20 ^{KF} +1 ^{KF})(monosomic)

↓↑

(2) Chinese (20 II^{Ch}) x F_1 (20^{CH}// 20^{KF} +1^{KF}) (monosomic)

↓ repeated back crosses (6-8) of this types will eliminate the chromosomes of KF except a chromosome which is monosomic and finally a F_1 would be

F_1 =(20 ^{IIch} + 1^{KF}) → selfing →(20 II ^{ch} +1 $^{II\,KF}$) = **substitution line**

4. Alien addition and substitution lines

The transfer of chromosomes/genes naturally occur in crop plants within the members of a particular species and sub-species or by interventions of crop breeders to genetically improve the crop plants. The simple procedure followed to achieve such an objective is to hybridize a recipient and donor parent, followed by back crossing of their F_1 for several generations to recurrent parent to restore its original genotype alongwith the new gene/s from the donor parent. However, if the recipient and donor parent are not members of the same species, this simple approach of transferring the desirable genes is some-what cumbersome. But on the other hand considerable genetic variability, particularly for biotic and abiotic stresses, exists in wild relatives or distantly related domesticated species which necessitates the transfer of such variability to crop plants.Therefore, to exploit this natural variability in recent times bridge crossing and embryo rescue technique are employed which have immensely facilitated the transferring of alien genes to crop plants. The transfer of variability is mainly carried out at genomic level (synthesis of *Triticales and Raphanobrassica*), at chromosomal level (production of addition and substitution lines, monosomics, trisomics and nullisomics etc and at chromosomal segment level. The introgression of a chromosomal segment has been achieved through reciprocal translocation or homoeologous pairing (introgression of *Ae. umbellulata* chromosome segment into bread wheat by sears and 1B/1R translocation). The treansfer of an alien chromosome with preferred genes from wild or distantly related source to a recipient agronomic genotype deficient in terms of a specific trait is known as **alien gene transfer**. In the present context only chromosome manipulation at chromosome level will be dealt particularly the cases of alien addition and alien substitution in bread wheat.

1. Scheme for alien substitution

♀ Monosomic recipient ($20^{II}+1^{I}$) x ♂ wheat alien addition line ($21^{II}+1^{II\ alien}$)

↓

F_1 → 20 II (wheat)+ $1^{Iw}+1^{I\ alien}$

↓ (selfing)

Some progeny ($20^{II\ wheat}+1^{II\ alien}$) = a substitution line

2. Scheme for Alien addition

The general procedure to develop alien addition line was suggested by Sears (1953). The addition of alien chromosomes without disturbing the genic balance of recipient variety is tolerated by polyploid species to an large extant while in case of diploids usually even small deficiencies and duplications are deleterious and hence development of alien addition lines is almost impossible. However, Yamashita et. al. (2007) in Japan reported the development of addition monosomic lines even in case of diploid onion based on the procedure given by Shigyo et al. (1997). The addition of alien chromosomes to a recipient variety involves the crossing of parent with higher chromosome number as female parent and with lower chromosome number as male parent. The F1 in such wide or distant crosses is usually sterile, hence to make it fertile its chromosome number is doubled by applying the colchicine. The amphiploid thus derived is repeatedly backcrossed for 6-8 generations to recover the original genotype of recurrent parent and eventually some plants are obtained which have original chromosomes of the recurrent parent plus an alien chromosome as monosome. These lines are called as monosomic alien addition lines MAALs. Which produces alien disomic lines after selfings (Fig 70).

(a) Wheat-rye addition line

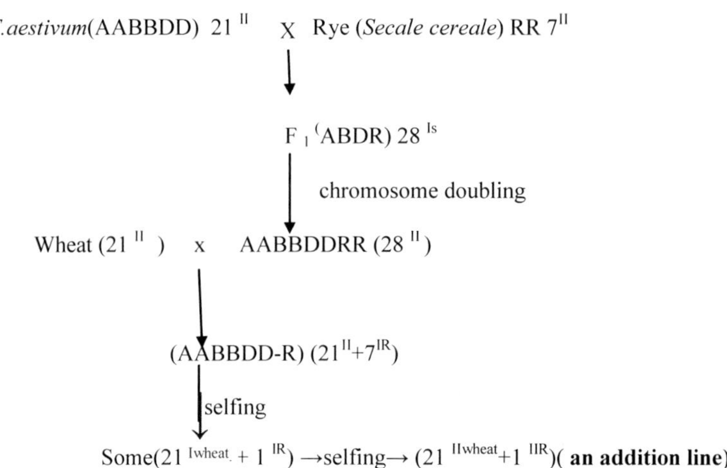

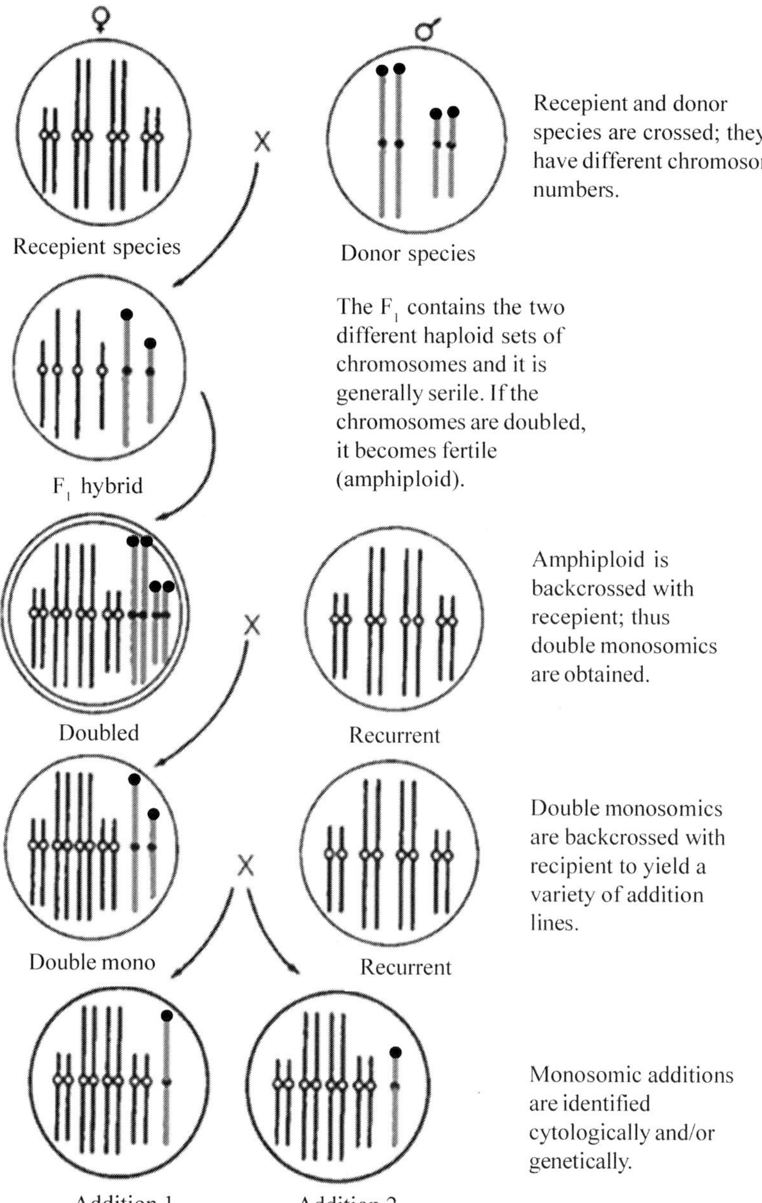

Fig. 70: Scheme to produce Monosomic addition lines.

(B) Procedure to develop alien addition line in diploid *Allium* (Shigyo *et al.* (1996)

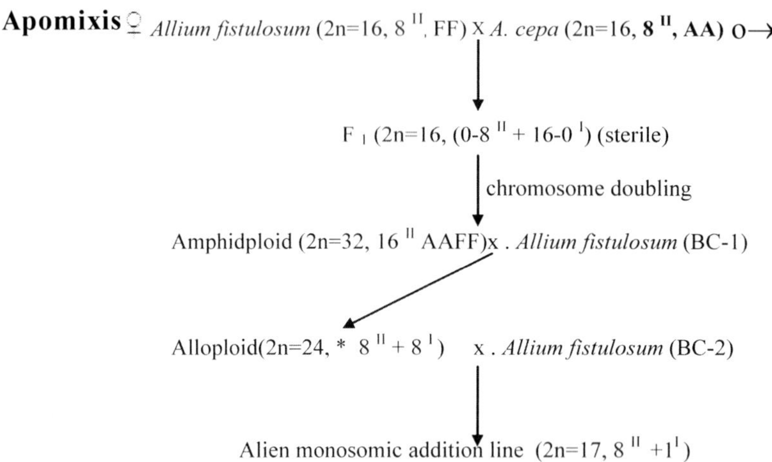

Apomixis is a combination of two Greek words, i e, apo (away from) and mixis (mixing or fusion) which simply means without mixing of maternal and patermal gametes and thus antonym to **amphimixis.** In other words, apomixis could be defined as the occurrence of a reproductive process other than the normal sexual processes which involve reduction of chromosome from 2n to n through meiosis and fertilization to restore the somatic (2n) chromosome number. Contrarily under apomictic mode of reproduction formation of seed occurs from the maternal tissues of the ovule by passing the processes of meiosis and fertilization, leading to embryo development. The initial discovery of apomixis in higher plants was attributed to the observation that a solitary female plant of ***Caelebogyne ilicifolia*** from Australia continued to form seeds when planted at Kiev Gardens in England (Smith,1841). Winkler (1908) first coined the term apomixis to mean" substitution of sexual reproduction by an asexual multiplication process that does not involve nuclear or cellular fusion". Hence some authers used term apomixis to describe all forms of asexual reproduction in plants which is no longer generally accepted and the current usage of apomixis is synonymous with the term "agamospermous, ie, production of viable embryos asexually"(Richards,1997). The apomictics are classified in to two main categories, ie obligate and facultative ones. The obligate ones reproduce only by apomixis while facultatives reproduce through both means of reproduction, ie gametic as well as apomictic one.

In normal sexual cycle developmental steps occur inside the ovule to produce the female gametophyte (embryo sac) followed by double fertilization to give rise embryo (2n) and endosperm (3n). The meiotic division converts the megaspore mother cell (mmc) into four spores (quartet) and out of these four spores three are degenerated and one develops into an 8- nucleate embryo sac by three succeeding mitotic divisions. When two sperm cells enter the embryo sac through micropylar end, one of the sperm cell fuses with egg and other one with central cell nulei to produce embryo (2n) and endosperm (3n), respectively. But under apomictic mode of reproduction the development of functional female gametophyte occurs without meiosis and double fertilization and embryo develops automatically from unreduced female gamete. The development of functional endosperm in some apomicts occurs autonomously while in pseudogamous apomicts fertilization of the central cell by sperm cell may be required. Actually apomixis or asexual reproduction occurs when sexual life cycle is "short circuited (Vielle-Calzada *et al.,* 1996). The life cycle of sexual angiosperms follows the alteration of generations, in which meiosis precedes the formation of gametes, and the double fertilization restores the somatic chromosome number. While short- circuiting of sexual cycle results in directly formation of seed having genotype of maternal parent (Fig. 71).

The apomixis is a complex phenopmenon having many controlling mechanisms. However, on the basis of some commonalities so far identified it could be categoried into two broad sub groups as ***gametophytic*** (diplospory and apospory) and **sporophytic** or adventitious embryony

1. **Gametophytic apomixis**: Gametophytic apomixis operates through mediation of an unreduced embryo sac where endosperm development occurs either spontaneously or induced by fertilization (pseudogamy). It is further subdivided into two categories,ie, **diplospory** and **apospory**, based on the cell type that give rise to unreduced embryo sac. In diplosporous types, the megaspore mother cell (MMC) or a cell with apomictic potential occupying its position is the progenitor cell for the unreduced embryo sac. The cell may enter meiosis but this aborts, and development proceeds by mitotic division to achieve embryo sac formation. It is called **meiotic diplospory** and in event of meiotic diplospory, one of the megaspores that constitute the dyad (rather than the linear tetrad of megaspores in sexual life cycle) degenerates. After three mitoses, an embryo sac of 8 nuclei results as usual having unreduced chromosome number of the parental sporophyte. Alternavely, that cell might under go directly to mitosis to form an reduced embryo sac, which is termed as **mitotic diplospory**. In both cases the eight nucleate embryo sac is formed by three mitotic divisions

   Under **aposporous** mode of apomixis, one or more somatic cells (nucellar cells) of the ovule, called aposporous initials, give rise an unreduced embryo sac. Aposporous initials exist most frequently adjacent to the cells under sexual reproduction, but they do not necessarily arise from nucellus, since epidermal and integumentary origins have also been reported (Koltunow and Grossniklaus,2003) Aposporous initials can differentiate at various times during ovule development. Meiotically induced and aposporous embryo sacs might co-exist in the ovule but aposporous embryo sac might continue to develop while sexual embryo sac degenerates.

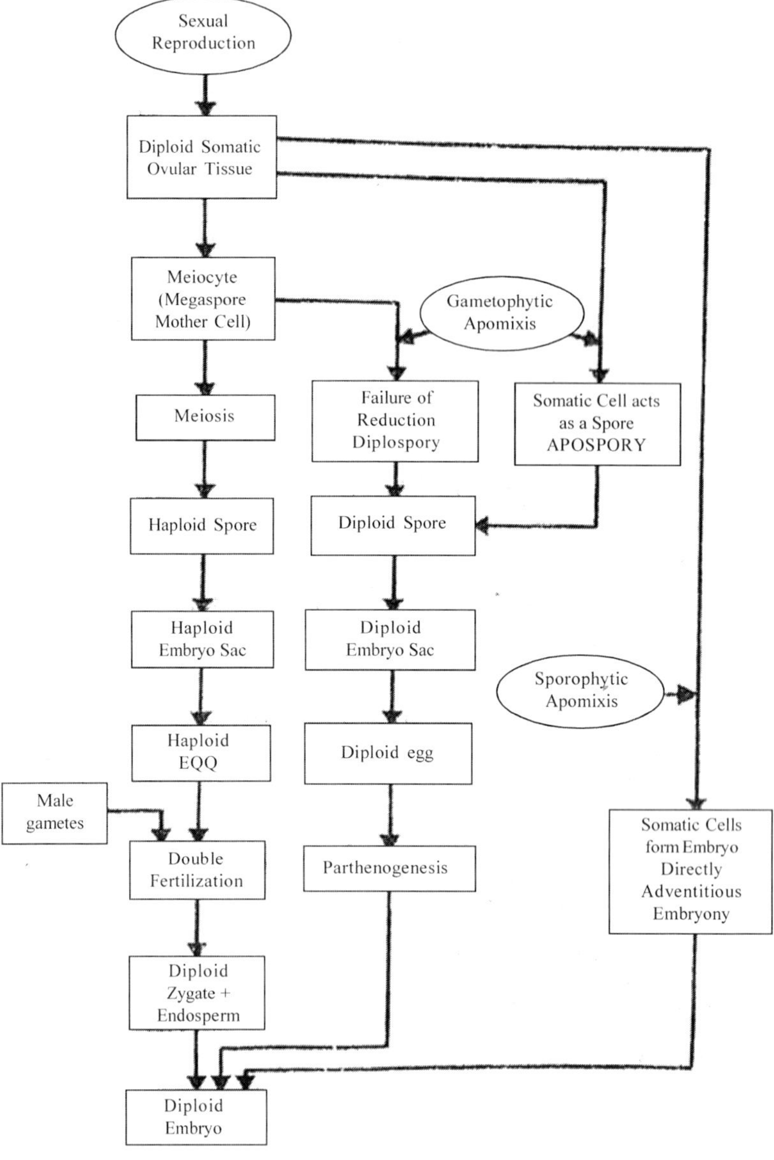

Fig. 71. Apomictic pathways (apospory, diplospory and and sporophytic), as a deviation from normal sexual reproduction.

2. **Sporophytic apomixis**: It is also referred to as adventitious embryony and in sporophytic or adventitious embryony, embryo directly arises from nucellus or the integument of the ovule (Koltunow et al. 1995). For development of viable seeds under normal fertilization in the adjoining embryo sac subsequent endosperm formation is necessary to provide nourishment but in case of sporophytic or adventitious embryony, parthenogenetically developing embryos closest to the embryo sac grow towards it presumably to obtain nutrient and other developmental signals from embryo sac.The bud like embryo development is rapid; and often multiple embryo form which hinder development of the zygotic embryo.

Genetic basis of apomixis

The first known study of inheritance in an apomictic plant was unknowingly conducted by Gregor Johann Mendel (1869) on *Hieracium*, an apomictic plant, to corroborate his laws of inheritance (Nogler,1994). In contrast to his observations in pea, the *Hieracium* F_1 hybrids showed extensive segregation and F_2 populations were consistently uniform without any segregation. Thus his observations in Pea and *Hieracium* were contradictory and hence, perhaps, Mendel abandoned *Hiercacium*. However, by the turn of 20th centuary, apomixis was a known phenomenon in plants. Ostenfeld (1904-1910) and Rosenberg, (1906-7) reported the expression of apomixis in genus *Hieracium*. Hanna et al. (1998) worked out the genetic basis of apomixis and reported it to be a qualitative trait controlled by few genes involving many intriguing aspects of genetic analysis. Since apomixis is not common in cultivated species, hence its genetic analysis was conducted in unknown polyploids and heterozygous wild plants. Sevidan (2000) reported an obvious association between gametophytic apomixis and polyploidy and reported apomixis to be caused by dosage effects of genes. However, presence of apomixis, a rare one in diploids indicates that polyploidy does not appear to be absolutely required for the expression of the apomixis.Genetic mechanisms like epistatic interactions between apomixis controlling genes, sporophytic and gametophytic factors, modifiers of gene expression, segregation distortion, suppressed recombination, presence of asymmetrical regions and environmental factors were all reported to be involved in expression of apomictic phenotype. Nevertheless,low fertility, interspecific crosses, genome irregularities, difficulties in observing apomixis and occurrence of both apomictic and sexual reproduction in the same plants hinder proper genetic analysis of apomixis. Further more, Ortiz et al. (2013) reported apomixis to be under epigenetic control and therefore, as a conclusion,it seems to be controlled by alternative mechanisms in different plant species.

Some other studies regarding genetic control of apomixis indicated it to be controlled by a single gene with some modifiers.Carneiro(2006) reported that

aposporous and **diplosporous** mechanisms in grasses to have simple inheritance controlled by a dominant locus.Savidan (1981) in *Panicum maximum* and Bicknell *et al.* (2000) in *Hieracium* also reported apomixis to be under control of simple dominant factor. The apomeiosis (suppressed meiosis and parthenogenesis (seed development without meiosis) were tightly linked and under **pleiotropic c**ontrol in some studies, while recent reports with the genera *Taraxacum, Poa, Allium, Hypericum, Hieracium, and Potentilla,*established them to be under independent control (Ozias-Akins and van Dijk, 2007).

However, all genetic mechganisms worked out so far are reported to be based upon the data obtained by crossing between closely related sexual and apomictic species where often apomictic was a pollen parent and reciprocal crosses were seldom attemped, hence maternal effects remained largely untested. Besides there is lack of proper identification of apomixis based on presence of genetically identical seedlings coupled with embryological study to confirm the reproductive mode of each progeny and assessment of ploidy in all individuals (Nogler, 1984a). Hence, it is speculated that a number of controlling loci have been under estimated, and rather than a single locus being involved, it appears that as many as three unlinked dominant factors may need to be inherited to ensure the inheritance of autonomous diplosporous apomixis in *Taraxacum officinale* (DIjk *et al.* (2003). Furthermore, as majority of the published data claimed apomixis to be qualitative trait, one that is either inherited or not, but in most plant it is expressed in facultative form where level of viable asexual seed formation can vary considerably between individuals and in F_1 the expression of apomixisis is below that of apomictic parent (Koltunow *et al.* (2000). The complexicity of its inheritance was also revealed in developing apomictic maize by Savidan and co-workers through introgression of *Tripsacum* DNA into Zea mays by repeated back crossing. But integration of intact functional apomixis into a maize genotype with just the 20 chromosomes of maize with small *Tripsacum* DNA remained a intractable objective.Since no fertile apomictic plants could be confirmed with fewer than 16 *Tripsacum* chromosomes (Savidan, 2001).

Hence inheritance of apomixis in not so simple as usually reported and some believe that individuals within a typical agamic complex, both sexual and apomictic, may share a predisposition to express this trait. The nature of predisposition may be a developmental characteristic,such as the possible presence of a large nutritive nucellus in Citrus or a nutritive integument in *Hieracium* (Koltunow *et al.*, 1995b and 1998) or may be genetic, such as the apparent linkage grouping in *Tripsacum* necessary for interspecific transfer (Savidan, 2001).

In most of the apomictic systems a small number of loci were reported to determine the inhereitance of the trait.A map based approach was attemped by Conner *et al.* (2015) who studied molecular mechanisms controlling apomictic mode of asexual reproduction leading to clonal seed formation in *Pennisetum squamulatum*. In gametophytic apomixis a chromosomally unreduced embryo sac develops from megaspore mother cell (diplospory) or from a near by nucellar cell (apospory) in process termed as **apomeiosis**. Parthenogenesis, is the development of unreduced unfertilized egg into an embryo constitute the second step of the apomictic process.In *Pennisetum squamulatum* apomixis was reported to be transmitted by a phycally large, hemizygous nonrecombining chromosomal region (ASGR). Multiple copies of the PsASGR-BABY-BOOM-like (PsASGR-BBML) gene reside within the ASGR and are postulated as strong candidate gene for the apomictic function of parthenogenesis based on linkage to the ASGR. A loss of the CcASGR–BBML in *Cenchrus ciliaris* ASGR recombinant plant was reported to have lost the ability to undergo parthenogenesis. In *Pennisetum squamulatum*, apomeiosis segregates as a single dominant locus, the apospory specific genomic region (ASGR).The ASGR contains multiple copies of the psASGR-BABY BOOM –like (psASGR-BBML) gene,a member of the BBM like sub group of APETALA2 transcription factors. Expression of psASGR-BBML transgene in sexual tetraploid pearl millet promoted parthenogenesis (embryo formation without fertilization) and production of haploid offspring. This study presents the first demonstration to our knowledge, of a function of a gene cloned from a naturally occurring apomictic plant that encodes a key component controlling parthenogenesis.

Detection and identification of apomixis in plants

The following indicators may show the presence of apomixis in higher plants:

1. Uniform progeny from heterozygous F_1 or open pollinated parents.
2. Maternal types in crosses.
3. High seed set in meiotically disturbed genotypes.
4. Multiple ovules seedlings per ovule
5. Seed setting in emasculated and bagged flower of an apomictic plant indicates presence of autonomous apomixis otherwise pseudogamous one.

Apomixis in crop improvement

Apomixis an asexual method of reproduction through seed provides unique opportunities for developing superior crop varieties. Apomixis occurs at low levels in some cultivated species and their closely related wild species. The transfer of apomixis can be exploited to evolve true breeding hybrids (without recombination or crossing over) and commercial production of hybrids without a need of male sterile systems. Since production and maintenance of sterility systems are highly cost oriented and labor intensive process. The obligate apomictic hybrids would breed true regardless of their heterozygosity and could provide a novel method for gene interogression in new genotypes. Obligate apomictics would be ready for performance testing without any progeny testing which normally requires 5-6 years to determine their stability and competitiveness in normal breeding programmes Therefore, apomixis has a great potential in crop improvement since it may facilitate multiplication of F_1 hybrid seed itself which is not possible through normal sexual reproduction. Normally hybrid seed is produced by hybridizing two inbred parental lines which are maintained as pure lines year after year. Their maintenance requires money as well as skilled persons which makes it highly expensive. Moreover, hybrid varieties (F_1) undergo segregation and recombination and hence their seed can not be used by farmer for the next-crop season. If apomictic mode of reproduction is installed in hybrid genotype, it would perpetuate hybrid vigour indfinitely due to absence of segregation and recombination without need of F1 seed every year from inbred parents. Apomixis would also remove the need of pollination for fruit and seed production. Extreme and adverse envirionmental condtion like drought and cold inhibit normal pollen development in crop plants inducing pollination failure and considerable yield loss. The impact of such environmental conditions could be minimized by apomictic crop production.

In interspecific or wide crosses, there is a tendency of chromosomal elimination and reverting to only one parent chromosomal complex. Apomixis may ensure chromosomes of both parents and survival of interspecific or wide crosses since chromosomal irregularities in meiosis and hybrid sterility would not be the norm under apomictic mode of reproduction.

Therefore, during the last decade sincere efforts are being made to introduce apomictic systems in crop plants. The first step to install apomixis in crop plants requires detection and identification of an apomictic plant source for its incorporation in a targeted crop artificially through hybridization between apomicts and amphimicts.Preferably the chromosome number in donor and recipient crop must be the same for easy transfer of apomixis to recipient plant.The other important factors are ploidy level and genome homoeology because apomixis is predominantly found in polyploids and genome homoeology facilitates recombination in donor and recipient parents.Although substantial evidence is available for hybrid origin of many of the apomicts but it is not necessary that all hybridization can induce apomixis (Stebbin,1950). The transfer of apomictic complexes is further complicated by breaking down of interacting gene complexes conditioning apomixis.The progress for incorporating apomixis in crop plants during recernt years has been summarized by Bicknell and Koltunow(2004). They reported that researchers are now in stronger position to install apomixis in crop plants and some basic issues like role of polyploidy in apomixisis and changes in the small number of key events of normal reproduction still remains to be completely resolved. Since understanding of these aspects might be exploited to identify and manipulate those elements to engineer asexual seed formation. The knowledge of cues and genes that enable cells in the ovule to switch to an apomictic pathway. The interaction between embryo and endosperm development in apomicts is poorly understood as is the capability of apomicts to tolerate parental genome imbalances during endosperm development that result in seed sterility in sexual plants. The role of epigenetic factors in control of apomixis is unknown yet they are widely implicated in control of sexual reproduction.

Nevertheless, there exists a possibility of transferring the apomixis in non-apomicticts. A research project during 1989 was initiated by CIMMYT, Mexico to install apomixis in maize (2n=20) from its wild relative *Tripsacum* (2n=72) by repeated back crossing. The results established a possibility to recover maize plants having 20 chromosome of maize and small DNA segments of *Tripsacum* (Fig.72).

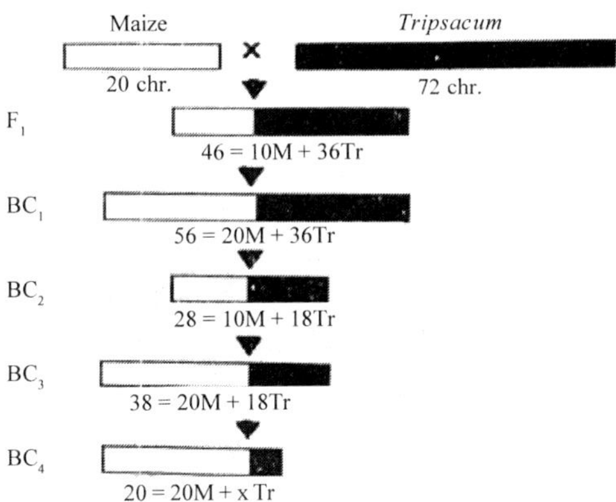

Fig.72 : Production of Maize-*Tripsacum* additition lines by four backcrosses (CIMMYT, 1983)

All the pros and cons of apomixis installing in crop plants were discussed by Bicknell and Koltunow (2004) who concluded that methodology of its installment has been comprehended considerably in recent times by researchers. However, some basic aspects as the role of polyploidy in apomixis and changes in the small number of key events of normal reproduction that lead to apomictic reproduction stillremain to be completely resolved. Since basic understanding of these aspects is an essential component to identify and manipulate the elements responsible for asexual seed formation crop plants.

References

1. Bicknell, R.A. and Koltunow, A.M. (2004). The Plant Cell 16:s228-s245 (supplement 2004).
2. Conner, J.A; Mookkan, M; Huo, H; Chae, K. and Peggy, O.A. (2015). Proc. Natl. Acad. Sci. U.S.A. Sep. 8: 112(36): 11205-11210.
3. Koltunow, A.M; Bicknell, R.A. and Chaudhury, A.M. (1995a). Plant Physiol. 108: 1345-1352.
4. Koltunow, A.M; Johnson, S.D. and Bicknell, R.A. (2000). Sex plant reprod. 1: 253-266.
5. Koltunow, A.M; Johnson, S.D. and Bicknell, R.A. (1998). Sex plant reprod. 11: 213-230.
6. Koltunow, A.M; Soltys, K; Nito, N. and Mc Clure, S. (1995b). Can. J. Bot, 73: 1567-1582.
7. Koltunow, A.M. and Grossniklaus, U. (2003). Annu. Rev. Plant Biol. 54: 547-574.
8. Koltunow, A.M; Sollys, K; Nito, N. and Mc Clure, S.(1995b). Can. J. Bot.73: 1567-1582.
9. Koltunow, A.M; Bicknell, R.A. and Chaudhury, A.M. (1995a). Plant Physiol. 108: 1345-1352.
10. Law, C.N. and Worland, A.J. (1996). Euphytica 89 (1): 1-10.
11. Nogler, G.A.(1984a). Gametophytic apomixis. In Embryology of Angiosperms, B.M. Johri, ed. (Berlin : Springer-Verlag), pp. 475-518.
12. Nogler, G.A. (1984b). Bot. Helv. 94: 411-422.
13. Nogler, G.A. (1994). Pol. Bot. Stud.8: 5-11.
14. Ortiz, J.P.A; Quarin, C.I; Pessino, S.C; Acuna, C; Martinez, E.J; Espinoza, F; Hozgaard D.H; Sarter, M.E; Caceres, M.E. and Pupilli, F. (2013). Ann. Bot. (London) 112 : 767-787.
15. Ostenfeld, C.H. (1904). Bot. Ges. 22: 537-541.
16. Ostenfeld, C.H. (1906). Bot. Tidsskr. 27: 225-248.
17. Ostenfeld, C.H. (1910). Z. Indukt. Abstammungs.Vererbungsl. 3: 241-285.
18. Ozias-Akins, P. and van Dijk, P.J. (2007). Ann. Rev. Genet. 41: 609-537.
19. Richards, A.J. (1997). Plant Breeding Systems, 2nd Ed. (London Chapman and Hall).
20. Rosenberg, O. (1906). Ber. Dtsch. Bot. Ges. 24: 157-161.
21. Rosenberg, O. (1907). Bot. Tidsskr. 28: 143-170.
22. Sears, E.R. (1953). Am. J. Bot. 40: 168.
23. Sevidon, Y. (2000). Plant Breed. Rev. 18: 13-85.

24. Sevidon, Y. (2001). Transfer of apomixis through wide crosses. In flowering of Apomixis: From mechanisms to Genetic Engineering. Y. Sevidan, J.G. Carman and T. Dresselhaus eds. (Mexico CIMMYT IRD, European commission DG VI), pp.153-167.

25. Shigyo, M; Tashiro,Y; Isshiki, S.and Miyazaki, S.(1997). Genes and Genetic Systems 71(6):363-371.

26. Smith, J. (1841). Notice of a plant which produces seeds without any apparent action of pollen.

27. Transactions of the Linnaean Society of London (meeting of June18, 1839). 18.

28. van Dijk, P.J; van Baarlen, P; and de Jong, J.H. (2003). Sex plant reprod 16: 71-76.

29. Vielle-Calzada, J.P; Nuccio, M.L; Budiman, M.A; Thomas, T.L; Burson, B.L; Hussey, M.A. and Wing, R.A. (1996). Plant Mol. Biol. 32: 1085-1092.

30. Winkler, H. (1908). Prog. Rei. Bot. 2: 293-454.

31. Yamasita, K; Masayoshi, S; Masuzaki, S; Yaguchi, S.Hang, T.T.M; Yamauchi, N.Endang, S; and Tashiro, Y. (2007). Genes, Genomes and Genomics. Global Science Books.

5

Polyploids: Reversion, Genome Restructuring and Alien Gene Transfer

The newly synthesized autopolyploids, having few or more undesirable features, e. g; meiotic instability, low seed setting and late maturity etc, are usually not suitable for immediate commercial use, are known as **raw polyploids**. This is specifically more applicable for those sexually propogated polyploids which are developed for seed. In early generations due to polyploid chromosome constitution and polysomic inheritance they predominanantly do not breed true. Further more, due to meiotic instability chromosome number in their progeny is not always constant or equal to as in synthesized autotetraploid which may vary from few more to few less than the standard autotetraploid. The raw polyploids usually reverts back to diploid parental constitution due chromosome elimination (Randolph and Fischer, 1939). The mechanism responsible for reverting back to diploid status is similar to that of which give rise haploids from diploids. Presently cases are known where either spontaneously or due to specific treatments the chromosome number was reduced to half in somatic tissues. The phenomenon causing such chromosomal reduction is termed as **somatic reduction** or **reduced mitosis** Earlier reports suggested it probably to be caused due to spindle organizer abnormalities or some other unknown mechanism/s. The treatment of seed or seedling by colchicine, chloramphenicol and para-flurophenylalanine are reported to have induced chromosome reduction and production of haploids.

The phenomenon of chromosomal elimination have been reported to occur in interspecific hybrids as well as in intergeneric or more distant hybrids (Kostoff 1943; and Kasha and Kao1970) . Uniparental chromosome elimination in wide hybrids is reported to be the result of different chromosome behavior of two parents. According to one hypothesis centromeres from two parents interact unequally with the mitotic spindle, leading to selective chromosome loss or elimination (Bennett et al., 1976), Finch 1983 and Mochida et al., 2004). Centromeres are the chromosomal loci that attach to spindle microtubules to mediate faithful inheritance of genome during cell division. They are epigenetically specified by incorporation of essential kinetochore protein CENH3 (CENP-A in humans or HTR12 in *Arabidopsis thaliana* (Earnshaw and Rothfield 1985; Talbert et al., 2002), a histone H3 variant that replaces conventional H3 in centromeric nucleosomes (Henikoff and Dalal 2005). The chromosomal location of CENH3 is assembly site for the kinetochore complex of active centromeres. The loss of CENH3 results in the failure of centromere formation and chromosomal segregation (Allshire and Karpen, 2008). The loss of CENH3 has been associated experimentally with occurrence of uniparental chromosome elimination (Sanei et al., 2011). The mechanism, of selective chromosome elimination during early development of *Hordeum vulgarew x H. bulbosum* embryos established the role of centromere specific histone CENH3 in process of chromosome elimination as: (1) centromere inactivity of *H. bulbosum* chromosomes triggers the mitosis-dependent process of uniparental chromosome elimination in *Hordeum vulgare x H. bulbosum* hybrids; (2) centromeric loss of CENH3 protein rather than uniparental silencing of CENH3 genes causes centromeric inactivity. They also proposed a possible model of how the mitosis–dependent process of uniparental chromosome elimination works in *Hordeum vulgarew x H. bulbosum* hybrid embryos. After fertilization, two parental CENH3 genes are transcriptionally active but CENH3 is only loaded into the centromere of *H. vulgare* and not on *H bulbosum* chromosomes, which may be due to cell cycle asynchrony of two parental genomes during mitotic G2 phase. It causes chromosome lagging in *H. bulbosum* due to centromeric inactivity during anaphase leading to micronuclei formation at quartet stage and thus chromatin degradation / elimination of *H. bulbosum.*

Allopolyploidy, alien gene introgression and Interspecific hybridization

Interspecific hybridization, a mating between individuals of two different species, facilitates to combine diverged genotypes into a nucleus. At natural level transfer of genetic information occurs only among members of a particular species only due to their interbreeding, but wide hybridization breaks such a barrier or limitation by allowing the transfer of genome of one species to another. It is very important for evolution and speciation of species alongwith since chromosome doubling in wide hybrids is responsible for origin of many allopolyploids. Repeated back crossing of wide hybrids to their parental species has also contributed to the evolution and speciation of some species by introgression of a chromosome or chromsomal segment from one species to another. The genes, linked with biotic and abiotic stresses available in related wild species, are being introgressed into crop plants to increase their capability in terms of these deficiencies. The manipulation of chromosomes in wide hybrids is carried out at either genomic level or at level of at single chromosome level or at a chromosomal segment level as given in the Fig. 73.

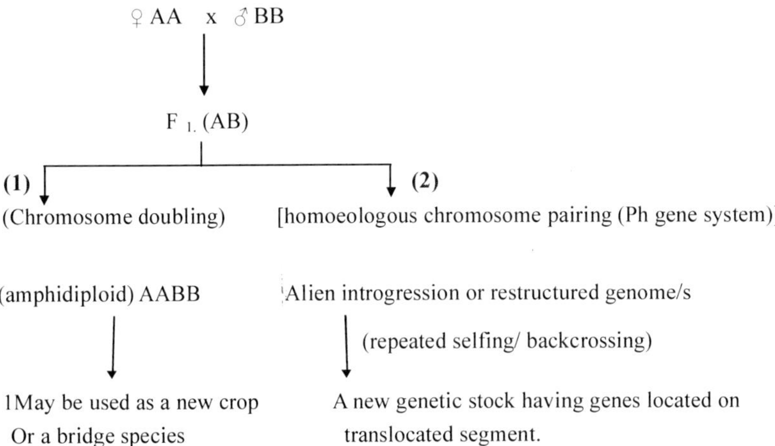

Fig. 73: The two unrelated genomes (AB) are thus brought together in a single nucleus by interspecific hybridization either may give a fertile progeny by doubling their chromosome number (amphidiploidy) **(1)** or may under go chromosomal rearrangement and diploidization to become a successful species in nature **(2)**.

1. Amphidiploidy

The F_1 hybrids obtained from distant hybridization (Interspcific/ intergeneric) are usually sterile since they have a single set of chromosomes from distant parents and so abnormal meiosis. To make them fertile ones having normal meiosis, the duplication of chromosomes is achieved. These fertility restored amphdiploids do have great importance in field of plant breeding to be used directly as a new variety, as a source to transfer genes from one species to another as a bridge species and to be used as a genetic stock having novel genes for crossing block. The doubling of chromosomes is usually carried out by application of highly toxic drug colchicine, extracted from a autumn *crocus* plant *(Colchicum autumnale)* . The drug inhibits spindle microtubule polymerization by binding to tubulin and thus blocking movement of chromosomes to poles during anaphase I. Spontaneous chromosome doubling have also been reported to occur in some interspecific and intergeneric hybrids resulting from unreduced gametogenesis and union of unreduced male and female gametes. Unreduced gametes with somatic chromosome number are believed to have played significant role in polyploidization and origination of both autopolyplods such as potato (Peloquin et al., 1999) and allopolyploids such as wheat (Kihara and Lilienfeld 1949). In recent times several workers forsee a great role of unreduced gametes in crop improvement such as a bridge to transfer alien genes into crop species and production of doubled haploids alongwith production of amphiploids (Ramana and Jacobson 2003; Zhang et al., 2007). The basic cause of origin of unreduced gametes is attributed to anomalies such as abnormal spindle orientation, defected synapsis and non segregation of chromosomes at one of the two meiotic divisions during meiotic cell division (Veilleux 1985; Ramana and Jacobson 2003). The production of unreduced gametes has been reported in amphiploid hybrids of tribe Triticeae. The haploidy–dependent unreductional meiotic cell division, resulting in unreduced gametes, has been considered a basic mechanism to have originated allopolyploid species in Triticeae.

Two cytological processes, ie, first division restitutution where lack of chromosome segregation occurs at anaphase I, followed by nuclear restitution and second meiotic division in hybrids between *T. turgidum* L; and *Ae. Tauschii* Coss (Xu and Joppa, 2000) and single division meiosis (SDM) characterized by mitosis like equation division with sister chromatids segregation at anaphase I are basic cause of unreduced gametes. However, the ultimate outcome of both types of chromosome segregation is same, ie, formation of genetically identical unreduced gametes called as mitosis like meiosis (Zhang et al., 2007).

2. Genome reorganization in polyploids

Another important aspect of polyploids, produced by wide hybridization, is genome reorganization. Modification of parental diploid genomes in polyploids was considered minimal, however, development and application of molecular techniques like chromosome painting, genetic mapping and comparative genetics have provided evidence for both intra and intergenomic reorganization of polyploid genomes. Chromosome painting has identified intergenomic chromosome rearrangements in polyploids relative to their diploid parents; nine such intergenomic translocations have been detected in allopolyploid tobacco *(Nicotiana)*, five intergenomic translocations in allotetraploid oats *(Avena maroccana)* and ~ 18 in allohexaploid *Avena sativa* (Leitch and Bennett (1997). A sound evidence of genome restructuring comes from comparative genetics of *Brassica.* Although gene order is conserved for million of years within plant families but genome restructuring also occurs rapidly and more extensively in polyploids than diploids (Gale and Devos, 1998) The analysis of synthetic polyploids and naturally occurring *Brassica* showed that naturally occurring allopolyploid genomes had extensive reorganization compared with their diploid progenitors. The analysis of *Brassica* also suggested two important points regarding genomic reorganization after polyploidization, (1) the more divergent the parents, the greater subsequent genomic change/s in the polyploid, and (2) the nuclear genome of maternal origin experiences less change than the paternal one Similar type of genomic changes do occur in allopolyploids of cereals also (Vega and Feldman, 1998). Presently availability of genomc maps in wheat (*Triticum*), barley *(Hordeum)*, rye *(Secale)*, oats *(Avena)* and maize *(Zea)* established the occurrence of chromosome rearrangements. In hexaploid wheat (*Triticum aestivum*) several intergenomic rearrangements, although less extensive than Brassica, are reported to have occurred after polyploididization,. Eilam et al. (2009) analyzing amount of DNA through cytometry in polyploids and their progenitors reported that additive value of genome size in typical autopolyploids. That nuclear DNA amount in cytologically diploidized autotetraploids exhibited considerable genome down sizing in *E elogatus* and instantaneous DNA sequence elimination in allpolyploid wheat group (Feldman et al., 1997).

Sometimes genomes present in allopolyploid are so drastically structured that they behave like diploids, the mechanism involved in such structuring is termed as **diploidization**. The diploidization is also reported to be caused by chromosomal reorganization and gene level changes (gene silencing / loss of duplicate gene expression) during evolutionary period. To elucidate the point further, the long standing debate about maize *(Zea mays)* can be cited, as to whether maize is a diploid or an ancient allopolyploid. The genetic mapping suggested it to be a case of an allopolyploid, having undergone extensive chromosomal rearrangement and presently documented as a partially diploidized or a degenerate polyploid (White and Doebley, 1999). The other well known and well documented example in this regard can be cited of bread wheat which inspite of having homoeologous chromosomes exhibit bivalent formation due to a gene located on $5B^{L}$ locus. These examples have established that chromosomal rearrangements and other genetic changes in genomes of polyploids played an important role in evolution. Further more, Matzke and Matzke (1998) reported that transposable elements (TEs) to have facilitated rapid genome restructuring after polyploidization in polyploids. They argued that polyploidy permits extensive gene modification by TEs because, by nature, polyploid genomes contain duplicate copies of all genes; hence they are well buffered from the deleterious consequences of transposition. The transposable elements do multiply and maintained in polyploids because of additional copies of genes they maintain which compensate for the loss of altered expression of genes that might result from TE insertion. The end result could be higher genomic restructuring in polyploids than their diploid progenitors. Transposable elements might also have been the driving force in the evolution of gene silencing mechanisms, such as methylation and heterochromatin formation in higher organisms (McDonald, 1998) The repression mechanisms might have evolved as adaptive responses to the selfish drive of TEs to expand in number in a host genome

The chromosomal and gene level changes during polyploidization represent a source of novel evolutionary processes. Rather than being stable, non interacting entities, two or more divergent genomes in a common polyploid nucleus could facilitate intergenomic interactions further leading to new chromosomal and gene arrangement and thus rtapid evolution. On the similar pattern genomic modification in autopolyploids which have complex polysomic inheritance does also occur which is termed as **allopolyploidization.**

Alien gene transfer (Wheat)

Wild species are a reservoir of untapped gene pool to improve the crop plants of economic importance. The introgression of a single chromosome or a chromosome fragment into a existing crop from their distantly related species, to improve their specific deficiency or to widen their genetic base is referred as alien gene introgression. The production of alien chromosome substitution, addition or translocation lines is basically aimed to transfer beneficial traits from wild species to cultivated ones. The transfer of a whole chromosome is usually not preferred since it might transfer some undesirable traits also (linkage drag) alongwith desirable targeted one/s. Hence majority of efforts attemped in this direction, aimed to transfer a chromosome segment having the preferred trait through alien chromosome translocation (Qi et al., 2007). The pioneer example in chromosome engineering in family Triticeae was attemped by Sears (1956) to transfer chromosome segment of *A. ubellulatum* having rust resistance to common wheat.

To transfer wild genes from wild species to cultivated ones, the first requirement is successful hybridizatioin between them. The success of wide hybridization primarily depends on the crossability of wheat genotype to be used in hybridization. and wild species. The crossability in wide hybridization is controlled by four crossability genes kr1, kr2, kr3 and kr4. The kr1, kr2 and kr3 are located on three homoeologous group 5 chromosomes, ie, kr1 (5B), kr2 (5A) and kr3 (5D) (Riley and Chapman 1967). These genes function in a complementary way where kr1kr1 loci contribute most to crossability frequencies. Some Chinese wheat genotypes are also reported to have a fourth loci kr4kr4 (Sears and Miller, 1985) . Chinese spring wheat contains recessive kr1, kr2, and kr3 and thus considered the best crossable wheat genotype. In earlier times wide hybridization usually failed due to improper development of endosperm but the advancement of embryo resque technique has solved this problem to an large extent. Under embryo resque technique, immature embryo is excised from 14-18 days post pollination and cultured on the artificial media for nourishment and proper development.

The wheat-alien translocations are achieved by three mechanisms:

1. Manpulation of 5B system or by inducing homoeologous chromosome pairing
2. Natural means
3. Ionizing irradiation

1. Manipulation of 5B system

To transfer chromosomal segment from wild relatives to cultivated ones, the mechanism of translocation is considered most promising one. Translocation may occur in terminal, intercalary or centric positions of chromosome. Centric or Robertsonnian translocation usually occur by the misdivision of univalents at meiosis giving rise to two telocentric chromosomes. Subsequent fusion of any two telocentric chromosomes of different origin leads to form a centric translocation. Un-paired chromosomes in wide hybrids are prone to misdivision and refusion. The homoeologous chromosomes of wheat and alien species show low pairing at meiotic metaphase I. due to presence homoeologous pairing inhibiting system (Ph) in wheat and thus preventing wheat-alien translocation. The inhibiting system consists of a major genePh1 located on $5B^{L}$ (Riley and Chapman1958 and Okamoto1957), an intermediate pairing gene, Ph2, on 3D (Mello-Sampayo 1971) and several minor loci (Sears1976). To induce wheat –alien translocation, the suppression of 5B system or promoting of homoeologous pairing is essential (Sears, 1972). The homoeologous pairing can be achieved either by using the mono-5B PhPh mutant or nulli-tetrasomic stocks as the maternal wheat sources. Recently development of wheat genotypes carrying the high pairing Ph^{1} gene derived from *Aegilops speltoides and Ae. mutica* also suppresses the effect of ph1 gene and promotes homoeologous pairing (Riley et al., 1958 and Kimber and Athwal, 1972). However, the efficiency of homoeologous chromosome pairing of Ph^{1} gene is not as high as those obtained by eliminating 5B (Chen et al., 1994).

The translocation induced by homoeologous pairing is reported to be directional manipulation since it occurs between alien chromosome and specific wheat chromosome, In such an experiment designed to induce translocations between *A. elongatum* chromosome 3Ae and wheat chromosome 3D, 17 out of 20 resulted translocations involved 3D and remaining three involoved 3B (Sears1972, 1976, 1982) Because the translocations involved homoeologous chromosomes of wheat and alien species where alien segment compensated for the missimg wheat segment. Nevertheless, chromosome arms containing genes involved in diploidization of wheat genome (3DS and 5BL), and containing fertility and stability genes, 2AS, 4BS and 6BS could not be replaced with alien segments.

Muzeeb-Kazi and Asiedu (1989) suggested scheme to obtain partially controlled translocation/s between homoeologous chromosomes of D genome of wheat and alien species to have derived from *T. aestivum* x diploid alien species and top crossed with *T. turgidum.* The derivatives possess 42 chromosomes if alien species is diploid., with a meiotic association of 14 bivalents (AABB) +7 univalents of the D genome+7 univalents of alien species. The 14 univalents of D genome and alien species provide basis for partially controlled translocations.

T aestivum (AABBDD)(2n=42) x Alien species (2n=14)

n=21 (ABD) ↓ n=7

F_1=21;ABD+7 alien x *T. turgidum* (AABB)(2n=28)

↓

Hybrid (AABB) 14^{IIs} +[7^{Is} D+ $7^{Is\,(alien)}$]

Or alternatively an alien disomic addition line can be developed and top crossed with *T. turgidum*, where a single alien univalent chromosome will have the opportunity to undergo random association via translocation with 7 univalents of the D genome of *T aestivum*.

T. aestivum (addition line) (21^{IIs}+$1^{II\,alien}$) x *T. turgidum* (2n=28;AABB)

n=(7A+7B+7D+$1^{I\,alien}$) ↓ n=7A+7B

Hybrid14^{IIs}AABB+ [7D+ 1^{alien}]

2. Naturally occurring wheat-alien translocations

The second group of wheat –alien translocation belongs to naturally occurring ones These translocations are reported to have significantly contributed towards conferring disease resistance. In this cotext examples of 1A/1R (1R confers resistance to several foliar diseases) and 1B/1R can be cited. The 1B/1R translocation has contributed significantly in respect of yields as well as rust resistance (Rajaram et al., 1983) 1R of rye confers resistance to several foliar diseases including genes for resistance to powdery mildew (Pm8 and pm17), yellow rustYr9), stem rust (Sr31) and leaf rust Lr26) along with wide adaptation and high yield potential. This translocation (1B/1R) presumably occurred from centromeric-breakage and reunion involving misdivision of univalent chromosome centromeres and reunion of the wheat and alien telocentric chromosome arms at the centromere. Further more the natural occurrence of 1B/1R translocation established a mechanism to artificially induce wheat-alien translocations (Sears1972). Other such translocations for an example, are as wheat-*Agropyron elongatum* translocation "Agent" which carries leaf rust resistance gene Lr 24 from *A. elongatum* (3Ae#1) involved exchange of similar size fragment of homoeologous chromosome 3D Smith et al., 1968), 5A/5R for copper deficiency and probable utilization of 6RL rye arm for cereal cyst nematode resistance. Genomic in situ hybridization (GISH) technique can be used to identify such wheat–alien translocations (Le et al., 1989).

3. Wheat-alien translocation induced by irradiation and other mechanisms

The third category of wheat- alien translocations are those which are produced either by ionizing irradiation (Sears 1956), by tissue culture (Lapitan et al., 1984) and application of gametocidal genes (Endo 1988). Gametocidal genes from several *Aegilops* species can also cause random chromosome breakage and then induce chromosome aberrations. The introduction of gametocidal genes, into alien addition/substitution lines, cause random wheat –alien translocations which can be recovered in selfed progenies. A large number of translocations have thus been produced but very few had some agronomic importance. Since most of the alien segments either do not compensate well for the lost wheat chromatin or contain undesirable linked genes.

Some reported examples of wheat-alien translocations in Triticeae are; "Transfer" of a segment from *A. umbellulata* chromosome 6U carrying Lr9 gene resistant gene for leaf rust to chromosome 6BL of wheat (Sears1956); stem rust resistanc from *A elongatum* to 6A of wheat (Knott1964); leaf rust and powdery mildew from rye to 4A of wheat (Driscoll and Jensen 1964); leaf rust and stem rust resistance from *A elogatum* to 7D of wheat (Sharma and Knott 1966); leaf-stem—and stripe rust resistance from *A. intermidium* to 7A of wheat (Weinhues 1973).

Ionizing irradiation induces at ramdom chromosomal breakage which has several advantages for wheat-alien translocations. Since target gene being located on the alien chromosome and has pairing potential with wheat chromosome without general efficiency being affected.

Secondly the alien chromosomes segment having target gene could possibly be inserted into a wheat chromosome without any chromatin loss of recipient chromosome. Nevertheless wheat-alien translocations are reported to have associated with some disadvantages also like genetic imbalance and deficiency for wheat segment and duplication of alien segment in relation to its wheat homoeologue. Due to these deficiencies few beneficial translocation, like wheat-*Aegilops* umbellulata (Sears 1956) and wheat - *A. elongatum* translocation (6AS-6AL-6Ae#1L) that carries stem rust resistance gene Sr26, have been realized. The Sr. 26 is reported to be a stable source of stem rust resistance and present in several Australian wheats. The success of wheat-alien translocations also depends upon the small size of alien segment which would not cause loss of wheat chromatin. Possibly such translocations induced by irradiation treatment are caused by two chromosomal breakages in alien chromosome having target gene and one breakage in recipient wheat chromosome. Mukai et al., 1992 reported such a intercalary translocation (4AS-4AL-6AL-4AL) having a resistance gene for Hessian fly derived from a rye chromosome 6RL.

Gene Transfer using amphidipoids as Bridge species

Bridge crossing is a strategic planning that exploits the reproductive superiority of polyploids to overcome the sexual incompartibilities existing between two species due to different ploidy levels. Under such circumstances, transitional or bridge crossing can be applied, followed by chromosome doubling to produce fertile bridge hybrids. To bypass the incompatible barrier between two genotypes/ species, a third species /genotype which is partly compatible with each of them. is used in immediate cross, for example, *Gossypium thurberi* (D_1 D_1) is incompatible with *G. hirsutum* (AADD) but can be crossed with *G. arboretum* (A_2 A_2). The amphiploid produced from *G. thurberi x G arboretum* can be hybridized with *G hirsutum*. Such a bridge cross in cotton was used to improve fiber strength of *G hirsutum* from the lintless *G. thurberi.*

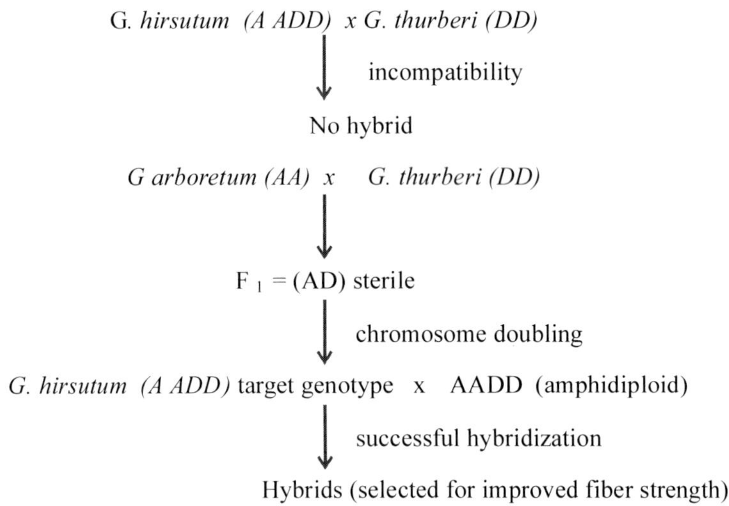

The methodology is usually applied where selection of the preferred trait is feasible. The desired trait incorporated from the wild species in target variety usually accompanied by several other undesirable characters also, therefore, several backcrosses are performed to eliminate them. Bridge crossing, for an example, was also employed by Sears (1956) to transfer a *Ae. umbellulata* chromosome segment to incorporate leaf rust resistance to bread wheat. Since direct crossing between bread wheat and *Ae. umbellulata* was not possible due to difference in ploidy level, therefore, first *T. dicoccoides* (AABB) (2n=28) a tetraploid wheat was crossed with *Ae umbellulata* (UU) (2n=14) ; Their progeny (ABU) (2n=21) being sterile was subjected to colchcine treatment for making it an amphidiploid (AABBUU and fertile). The AABBUU progeny facilitated transferring a UU chromosome segment to wheat. (Fig. 51). A third example of bridge cross can be cited of development of superior tall fescue grass (F. *arundinacea*) from Italian ryegrass (2n=2x=14) and tall fescue (2n=6x=42) by using meadow grass *(F. pratensis*) as a bridge species (Acquaah, 2007).

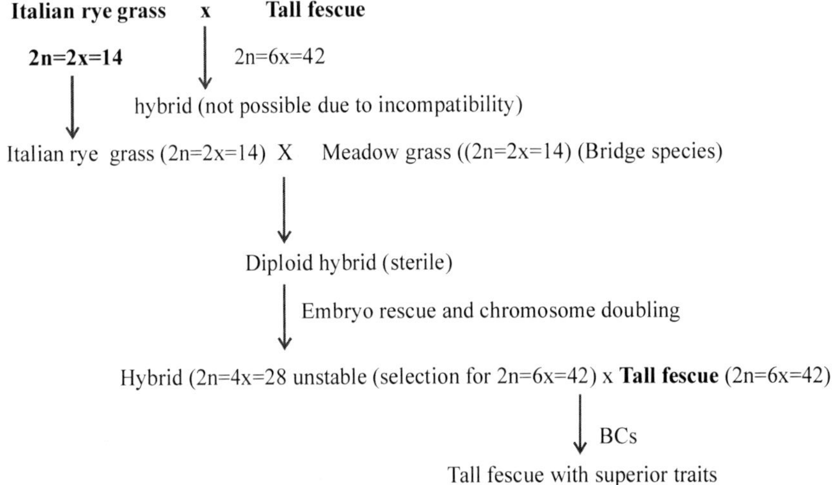

References

1. Alshire, R. C. and Karpen, G. H. (2008). Nat. Rev. Genet. 9: 923-937.
2. Bennett, M. D; Finch, R. A. and Barclay, I. R. (1976). Chromosoma 54: 175-200.
3. Chen, P. D;Tsujimoto, H. and Gill, B. S. (1994). Theor. Appl. Genet. 88: 97-101.
4. Driscoll, C. J. and Jensen, N. F. (1964). Crop Sci. 4: 372-374.
5. Earnshaw, W. C. and Rothfield, N. (1985). Chromosoma 91: 313-321.
6. Eilam, T; Anikster, Y; Millet, E; Manisterski, J. and Feldman, M. (2009). Genome 52(3): 275-285.
7. Endo, T. R. (1988). In: Miller, T. E. and Koebner, R. M. D. (eds.) Proc. 7th Int. wheat Genet. Symp. Cambridge, UK. pp. 259-265.
8. Feldman, M; Liu, B; Sega, G;Abbo, S; Levy, A. A. and Vega, J. M. (1997). Genetics 147: 1381-1387.
9. Finch, R. A. (1983). Chromosoma 88: 386-393.
10. Gale, M. D. and Davos, K. M. (1998). Science 282: 656-659.
11. Kasha, K. J. and Kao, K. N. (1970). Nature (Lond.) 225: 874-876.
12. Kihara, H. and Lilienfeld, F. (1949) Hereditas suppl: 307-319.
13. Kimber, G. and Athwal, R. S. (1972). Proc. Natl. Acad. Sci. U. S. 69: 912-915.
14. Knott, D. R. (1964). Can. J. Genet. Cytol. 6: 500-507.
15. Kostoff, , D. (1943). Der Zuchter 15: 121-125.
16. Lapitan, N. L. V; Sears, R. G. and Gill, B. S. (1984). Theor. Appl. Genet. 68: 547-554.
17. Le, H. T; Armstrong, K. C. and Miki, B. (1989). Plant Mol. Biol. Rep. 7150-158.

18. Leitch, U. and Bennett, M. D. (1997). Trends Plant Sci. 2: 470-476.

19. Matzke, A. J. M. and Matzke, M. A. (1998). Curr. Opin. Plant Biol. 1: 142-148.

20. Mc Donald, J. H. (1998). Mol. Biol. Evol. 15: 377-384.

21. Mello-Sampayo, T. (1971). Nature New Biol. 230: 22-23.

22. Mochida, K; Sasakuma, T. and Tsujimoto, H. (2004). Genome 47: 199-205.

23. Mukai, Y; Friebe, B; Hatchett, J. H. and Gill, B. S. (1992). In: Proc. 2nd Int. Symp. Chromosome Engineering in Plants Columbia M.O.

24. Muzeeb-Kazi, A. and Asiedu, R. (1989). In: Strengthening Collaboration in biotechnology: international agriculture research and private sector, USAID conf., Virg pp. 211-231.

25. Okomoto, M. (1957). Wheat Inform. Serv. 6: 3-4.

26. Peloquin, S. J; Boiteux, L. S. and Carputo, D. (1999). Genetics 153: 1493-1499.

27. Qi., L; Friebe, B; Zhang, P. and Gill, B. S. (2007). Chromosome Res. 15:3-19.

28. Rajaram, S. C; Mann, E. G; Ortiz-Ferrara and Muzeeb-Kazi-A. (1983). In Proc. Int. Wheat Genetics Symposium. 6th Kyoto, Japan.

29. Ramana, M. S. and Jacobsen, E. (2003). Euphytica 133: 3-18.

30. Randolf, L. F. and Fischer, H. E. (1939). Proc. Natl. Acad. Sci. Wash. 25: 161-164.

31. Riley, R. and Chapman, V. (1967). Genet. Res. 9: 259-267.

32. Riley, R. and Chapman, V. (1958). Nature 182: 713-715.

33. Riley, R; Unrau, J. and Chapman, V. (1958). J. Hered. 49: 91-98.

34. Sanei, M; Pickering, R; Kumke, K; Nasuda, S. and Houben, A. (2011). Proc. Natl. Acad. Sci. U. S. A. 108: 13373-13374.

35. Sears, E. R. and Miller, T. E. (1985). Cer. Res. Comm. 132: 261-263.

36. Sears, E. R. (1956). Brookhaven Symp. Biol. 9: 1-22.

37. Sears, E. R. (1972). Stadler Genet. Symp. 4, Univ. Missouri, Columbia pp. 23-38.

38. Sears, E. R. (1976). Annu. Rev. Genet. 10: 31-51.

39. Sears, E. R. (1982). Can. J. Genet. Cytol. 24: 715-719.

40. Sharma, D. and Knott, D. R. (1966). Can. J. Genet. Cytol. 8: 137-143.

41. Smith, E. L; Schlehuber, A. M;Young, H. C. and Edwards, L. H. (1968). Crop Sci. 8: 511-512.

42. Talbert, P.B.; Masuelli, R.; Tyagi, A. P; Comai, L. and Henikoff, S. (2002). Plant cell 14: 1053-1066.

43. Vega, J. M. and Feldman, M. (1998). Genetics 150: 1199-1208.

44. Veilleux, R. (1985). Plant Breed. Rev. 3: 252-288.

45. White, S. and Doebley, J. (1999). Genetics 153: 1455-1462.

46. Wienhues, A. (1973). In: Sears, E. R. Sears, L. M. S. (eds.) Proc. 4th Int. Wheat Genet. Symp. University of Missouri Columbia M. O; pp. 201-207.

47. Xu, S. J. and Joppa, L. R. (2000). Plant Breed. 119: 233-241.

48. Zhang, L. Q; Yen, Y; Zheng, Y. L. and Liyu, D. C. (2007). Sex Plant Reprod. 20: 159-166.

6

Pre and Post Fertilization Barriers in Crop Plants During Wide Hybridization and Techniques to Overcome

Wide hybridization (interspecific/intergeneric) is basically attemped to introduce new genetic variability from wild or distantly related species into cultivated ones or to synthesize new plant type/s like *Triticale and Raphanobrassica*. Majority of the commercial crop plants are usually deficient for one or the other desirable traits which are usually available in their wild relatives. However, crossability between commercial varieties and their wild relatives is very limited which is naturally imposed by certain genetically controlled reproductive barriers. These reproductive barrier/s usually operate either at pre or post fertilization stage/s in plant species (Stebbins1958). Since a species is constituted of individuals who share a common gene pool, reproduce among themselves only / exchange genes and share a common habitat. Hence at natural level gene flow remains strictly confined only among members of a particular species otherwise among the members of unrelated species a genetic barrier exists to create reproductive isolation. Therefore, wide crossing basically requires the overcoming of such barriers whch are imposed by temporal and spatial isolation of parents. The reproductive isolation consists of premating and post mating isolation mechanisms like habitat isolation or geographical isolation, sexual isolation and mechanical barrier depending upon incompatibility or mismatching of ganitalia (animals). Sexual isolation in animal kingdom is usually enforced by type of sexual displays and mating calls which are attractivce only to particular members of a species , however other potential intermating species also do exist in the same habitat or locality at that particular period of time.

Pre and post fertilization barriers in plants

The situations and events that occur prior to fertilization and block process of proper fertilization constitute prefertilization barriers. These are specifically relarted to stigmatic and styler barriers in plants. The overall situations preventing the interspecific hybridization may be as difference of flowering time between species, varying stigma receptivity and pollen viability, slow growth or no growth of pollen tube to reach to ovule and cross incompatibility or incongruity. In wide hybridization, once the fertilization has occurred, normal embryo growth may be restricted by post fertilization barriers. Since embryo and endosperm have to develop in equilibrium for sharing nutrients in an undisturbed developmental process. In general the first division of the zygote is delayed to favor the first division cycles of the endosperm cells to avoid disturbances in the equilibrium of development process of zygote and endosperm which cause abortion of young embryo and degeneration of endosperm. The abortion of young embryo may occur at various stages of its developmental stages in wide crosses. Therefore, embryo rescue, to be applied for normal development of embryo in vitro, may also vary accordingly to the requirement of these specific stages.

Methods to overcome barriers

1. **Crossability/genetic barrier**: The crossability between two species is controlled by genetic as well as environmental factors as for an example the pollen tube inhibition in interspecific crosses between sorghum and maize is reported to be controlled by a single gene (Laurie and Bennett(1989) and in some other species lethal gene/s are present which might induce early zygote degeneration in hybrid as in case of *Aegilops umbellulata* a lethal gene with three allelic variants does exist, ie, L^e for early lethality, L^l for late lethality and l for non lethality. Hence to obtain hybrids between (*Ae. umbellulata* x wheat (2x) , its non lethal strain should be used for hybridization. Similarly the crossability between wheat and rye (Intergeneric cross) is governed by Kr genes. The genotypes (Chinese spring wheat) having recessive (kr) alleles are more efficient to provide better seed setting when hybridized with rye. In general crossability among species depends upon the degree of genetic distance between them hence closely related species could easily be hybridized among themselves than distantly related ones. Therefore, prior to initiation of distant hybridization, the genetic architecture of parental genotypes should be well known. The degree of crossability may also vary among the existing accessions of a particular species due to their genotypic constitution. Hence all accessions available in germplasm of both the parents should be attempted in crossing programme. Moreover sometimes unidirectional crosses are more successful than reciprocal ones. Apart from use of crossable accessions some other alternative novel techniques may also be applied asVan Creij et al. (1993) suggested cut–style pollination technique in Lily to overcome such a genetic barrier.
2. **The use of mentor or mixed pollen**: To overcome the inhibition of growth of pollen tube in the style in many plant species the use of mentor/mixed pollen is used. The mentor pollen is rendered genetically ineffective by several different treatments including the action of ionizing radiation, washing in

solvents, or repeated cycles of freezing and thawing. However, all these pollen treatments are used but most commonly preferred method is gamma radiation. Despite of all these treatments treated pollens remain capable to germinate. The mentor pollen or mixed pollen technique is employed to overcome incompatible barriers by exploiting the fertilizing ability of the compatible pollen. In experiments mentor pollen is mixed with incompatible pollens of another species for pollination. The role of compatible mentor pollen is only limited to provide the stimulation for germination and successful fertilization by incompatible pollen.

3. **Manipulation of style and ovary for pollen tube growth:** In case of inhibition or slow pollen tube growth and lack of receptivity of stigma successful pollination can not be carried out on stigma. Therefore, alternatively, the pollination is carried out on placenta or ovules. Several reports are available in several plant species where manipulations of style and stigma have achieved smooth pollination. In some cases facilitation of normal pollen tube growth for successful feritilization involves removal of the stigma and a part or whole of the style and pollinating the cut end. The technique is known as **stump pollination** or **cut style** pollination. Since pollen tube growth either is inhibited just below the stigma or half way in the style. These both types of inhibitions can be overcome by cut-style manipulation (Van Tuyl et al., 1991; Janson et al., 1993). The cut-style manipulation facilitates the normal growth of pollen tube to ovary but it is reported to be associated with low seed set probably caused by premature arrival of pollen tube in the ovary (Janson et al., 1993) Alternatively Van Tuyl et al. (1991) suggested stylar graft technique to improve the seed setting in interspecific crosses. In this technique the pollen grains are first placed on the compatible stigma to initiate the pollen tube growth.. After one day the style of the pollen donor is excised 1-2 mm above the ovary and properly grafted on the ovary of another desired plant.

4. **Application of chemicals.** To improve seed setting after pollination in intersprcific crosses, some chemicals like auxins, cytokinins and gibberellins can also be applied to the ovary or to the pedicel at the time or just after the pollination. The application of these growth regulators causes delayed abscission of the style and development of young fruits (Alonso and Kimber1980; Muzeeb–Kazi 1981). The application of organic solvents like hexane and ethyl acetate before pollination is also reported to have overcome pre-fertilization barriers in Populus interspecific hybridization. (Willing and Pryor1976). For successful wide hybridization in cereals and Legumes the female parent could also be treated with immunosuppressors such as amino-n-caproic acid, salicyclic acid and acriflavin before or after the pollination.

Post fertilization barriers and techniques to overcome

During wide or distant hybridization usually the abortion of embryo occurs without any possibility of mature seed formation. To overcome this problem immature embryos are excised and cultured on artificial nutrient media which direcly develop into young seedlings. The technique involved is referred as embryo rescue which encompasses culturing of immature embryo, ovule and ovary on artificial media..

1. **Ovary culture**. When abortion occurs at an very early stage and maternal tissue does not affect negatively the development of seeds, ovary culture can be applied. It can also be applied when embryo culture and ovule culture are not possible due to very small size of ovules. In case of wide hybrids the ovaries are harvested 7-40 days after intrastylar pollination, surface sterilized and sliced into 2mm thick disks before placing on the media. To promote proper development of fruit, the chemicals like IAA, or coconut milk can also be added to the medium. Ovary culture has been applied in many species like *Brassica, Eruca-Brassica* hybrids , *Lilium* and *Phaseolus*. In some crops where ovary is latge enough to be cut into small parts and to be placed on media the ovary slice culture is preferred.

2. **Ovule culture:** When the development of the embryo and endosoperm is mismatched at an early stage and ovary culture is not feasible , the ovules can be dissected out after fertilization from ovaries and cultured on media to obtain mature seeds. The production of interspecific hybrids through ovule culture in *Alstroemeria* was reported by De Jeu et al. (1992). In *Alstroemeria* fertilization occurred after 24 hours of pollination, the ovaries were dissected two days after pollination and placed on the media containing 9% sucrose. After six weeks the ovules were placed on medium containing 4-5 % sucrose which started to germinate after 1-2weeks. However, the percentage of seedlings obtained was dependent of genotypic combinations in ovule cultures. Ovule culture was applied to produce hybrid seed in distant crosses of *Lycopersocon, Nicotiana*, and *Alstroemeria.*

3. **Embryo culture :** The embryo culture for the first time was employed by Hanning(1904) and later on was applied in several crops specifically for attempting interspecific and intergeneric crosses within the tribe Triticeae and Brassicaceae by several workers. (Williams et al. 1987). For culturing the embryos, the young embroys are excised under aseptic condition usually under laminar flow cabinet from ovules and cultured on a nutrient medium to provide suitable environment for their growth. Under modified embryo culture method (Harberd1969 and Buitendijk et al. 1992) the ovules are cut into half and the halves containting the embryo are cultured on liquid medium. Thereafter the halves possessing germinating embryos are placed on the solid medium under suitable temperature, photoperiod and humidity to obtain

plantlets. To produce interspecific and intergeneric hybrids in Brassicaceae sequential culture of ovary, ovule and often embryo has been applied by Nanda kumar et al. (1988) and Gundimeda et al. (1992). In sequential culture, ovaries are first cultured for 6-10 days and their enlaged ovules are excised to be recultured on fresh medium. The hybrid plantlets. may either be achieved from cultured ovules or embryos.

Prokaryotic chromosome (bacterial)

A prokaryotic *E. coli* cell is about 1000 times smaller than that of an average dividing plant cell. The prokaryotes have simpler structures as compared to eukaryotes, ie, plants and animals. The bacterial cells are enclosed by a rigid cell wall composed of peptidoglycan, a protein sugar polysaccharide molecule. The cell wall surrounds the cytoplsonic membrane, protects it from ambient environment and gives it a shape. The wall contains small appendages (fimbriae) and one of the fimbriae becomes pilus to facilitate conjugation (Fig. 73A). All bacterial cells are capable to give rise new daughter cells by fission and transmission of encoded genetic information possessed by them from one generation to another. The bacterial DNA is packaged in loops back and forth like a bundle and this bundled DNA, usually concentrated in a specific region of the cytoplasm called as **nucleoid** (Fig. 73B). The bacterial chromosome or genophore is a single (monoploid), circular, double stranded DNA molecule mostly attached to the plasma membrane at one point. The prokaryotic DNA basically differs from eukaryotic chromosomal DNA that neither it remains associated with histone, nor does it form nucleosome structures and nor does it have introns. The DNA of *E. coli*, a gut bacteria, is a circular, covalently sealed, molecule of 4. 6 million base pairs in length, having 4288 annotated protein-coding genes(2584 operons), seven ribosomal (rRNA) operons and 86 tRNA genes. However, certain bacteria like the *Borrelia burgdorferi* possess array of linear chromosomes like eukaryotes.

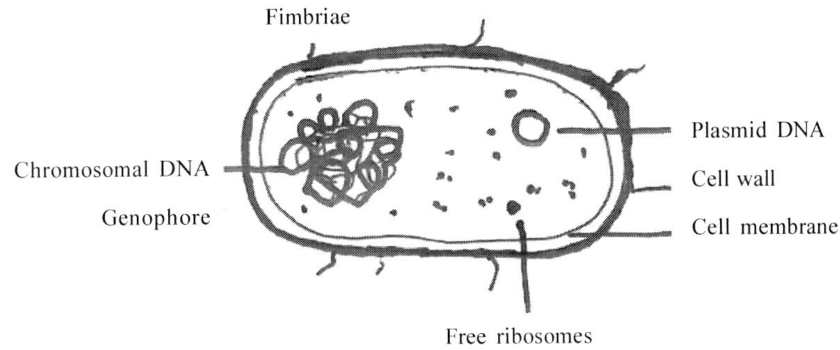

Fig. 73A: Structural organization of an *E. coli cell.*

Bacteria also contain genetic material besides the nucleiod. This double stranded DNA, also covalently closed and supercoiled, takes the form of plasmid (a circular DNA) which may amount to 0.5 to 2% of the cellular DNA. A plasmid may contain 4 μm upto 35 μm of duplex DNA and this too is free from histone. Some plasmid integrate into the host genome and are known as episomes. The plasmids encode products required for their replication and maintenance in the host. Most bacterial plasmids also carry genes for antibiotic resistance The plasmids remain floated in the cytoplasm alonwith ribosomes the only type of organelle found within the bacterial cell.

The single bacterial chromosome exists as a nucleo-protein complex which does not condense prior to the cell division as in case of eukaryotes. The strands of circular chromosome are so twisted at both ends that instead of forming a single circle, it is organized into a series of looped domains (Fig. 73B) which reflect functional as well as structural order. The looped domains of *E. coli* are approximately 50 kbp in length and approximately 100 in number in whole genome. Each domain appears to be isolated possibly to facilitate control over different promoters. However, molecular basis of this organization is almost uncharacterized specially in terms of that whether the loops define specific regions of the chromosome or not. It is also not obvious that what sequences of the chromosome are involved and how the loops are formed and maintained.

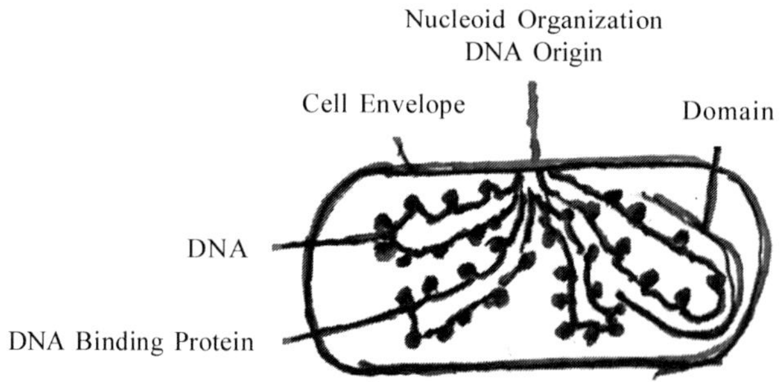

Fig. 73B : Structural organization of prokaryotic nucleoid.

Viral structure, genome and replication

The viruses are small, non cellular parasitic orgsanisms lacking all essential genetic information required for perpetuation of species. They intrinsically lack metabolic machinery and hence depend upon host cell for raw material and enzymes for essential functions such as replication, transcription and protein synthyesis. The viral genome is either DNA or RNA associated with proteins usually in the form of a shell or **capsid** without any organelles. Capsid, the protein coat of virus, is formed from a number of individual protein molecules called **capsomeres** which are arranged in a precise and highly repeatitive pattern around the DNA (genetic material). Several chemically distinct or a single type of capsomere may constitute the capsid. There are several types of capsids: the classic icosahedral capsid, spherical and bacilliform capsids and helical capsids.

The nucleic acid portion of the viruses is known as the **genome**. The nucleic acid may be double stranded, single stranded, linear, closed loops or in segments in viruses. The bacterial viruses (T phages,$\emptyset$ $80, P21$) have double stranded DNA while$\emptyset$ $X174$ and M13 have single stranded DNA. The bacterial viruses MS2 and R17 have single stranded RNA. The genetic marerial of plant viruses with few exception is predominantly RNA. The cauliflower mosaic virus , potato leaf-roll virus and some others have, however, double stranded DNA. The large rice dwarf virus (infects cereals and grasses) and the wound virus (infects legumes cabbages and clovers) have double stranded RNA. as their genetic material.

At structure level viruses show a considerable variation in shape and size. The appearance of most of the bacteriophages is like a tadpole (Fig. 74) having their genetic material in head like structure called **capsid.** Their tail has a inner core

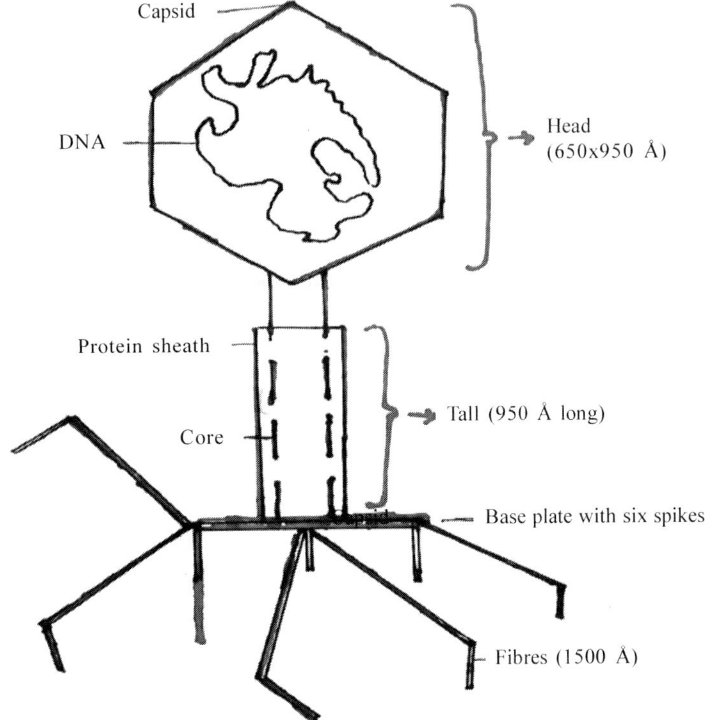

Fig. 74 : Structural organization of T-even bacteriophage (M. W≅. 2. 5x10 ^{8})

covered by a sheath. The tail end consists a base plate with spikes which help to anchor the bateriophage with host cell wall during infection. The T-even phages possess six fibers while the lambda phage has only one. The contact between cell wall of the host and bacteriophage is established by degrading/perforating the host cell wall by lysozyme like enzyme so that viral genetic material could enter the host. During this process viral protein coat is left outside the bacterium and only their genetic material is injected into the cytoplasm. However, plant viruses are required to cross the impermeable cuticle and plant cell wall before entering the cytoplasm. Hence viral infection in plants is often mediated mechanically by transfer from one damaged cell to another or via insect vectors. Many plant viruses encode proteins which interact with the plasmodesmata and allow semicomplete virions to spread from cell to cell.

With the introduction of the viral genome into host, the initiation of infection cycle begins and consequences of the infection vary according to the type of virus and host cell. In many cases, viral infection kills the bacterial cell – this is known as **lytic infection** and the phages causing such a infection are called as **virulent.** However Initial infection caused by **virulent** phages is followed by a latent perid where the virus multiply, then release of the progeny virus from the cell occurs by lysis. Alternatively, the virus may co-exist with the cell and produce progeny virus from the cell by continued budding or extrusion. The infection type is known as **persistent**. Many bacteriophages are capable of latent infection where virus is maintained in the host cell but does not produce infectious progeny which are called **temperate**. They lysogenize their host but also can undergo lysis if they re-enters the lytic cycle (induction). Many temperate phages lysogenize their host by integrating into the bacterial genophore which is described as **prophage**. Some eukaryotic viruses can also integrate with normal chromosome and are termed as **proviruses**.

During viral replication, as stated before the protein capsid is left on the host surface, the virus commands the host cell to synthesize components of new viral particles to be assembled to create new viral progeny. In case of viral genome consisted of RNA, it would directly assume the role of mRNA for synthesis of enzyme to be utilized for synthesis of viral genome and components of capsid to be utilized for further assembly of new viruses. In case the genetic material of virus is DNA, its genetic code is used for synthsis mRNA and rest of the process proceeds as usual.

References

1. Alonso, L. C. and Kimber, G. (1980). Cereal Res. Commun. 8: 355-358.
2. Buitendijk, J. H; Ramanna, M. S. and Jacobson, E. (1992). Acta Hortic. 325: 493-498.

3. De Jeu, M. J; Sasbrink, H; Garriga, Caldere, F. and Piket, J. (1992). Acta Hortic. 325-571-575.

4. Gundimeda, H. R;Prakash, S. and Shivanna, K. R. (1992). Theor. Appl. Genet. 83: 655-662.

5. Hannig, E. (1904). Z. Bot. 62: 45-80.

6. Harberd, A. (1969). Euphytica 18: 425-429.

7. Jansen, J; Reinders, M. C; VanTuyl, J. M. and Keijzer, C. J. (1993). Acta Botanica Neerlandica 42(L4): 461-472.

8. Laurie, D.A. and Bennett, M. D. (1989). Annals of Botany 64: 675-681.

9. Muzeeb-Kazi, A. (1981). J. Hered. 72: 227-228.

10. Nandakumar, P. B. A; Prakash, S. and Shivanna, K. R. (1988). In: Sexual Reproduction in Higher Plants (eds.) M. Cresti, P. Gori and E. Pacini, pp. 95-100. Berlin Heidelberg, New York: Springer Verlag.

11. Stebins, G. L. (1958). Adv. Genet. 9: 147-215.

12. Van Creij, M. G. M; Van Raamsdonk, L. W. D. and Tuyl, J. M. (1993). The Lily yearbook of the North American Lily Society 43: 28-37.

13. Van Tuyl, J. M;Van Dien, M. P ; Van Creij, M. G. M ; Van Kleinwee, T. C. M; Franken, J. and Bino, R. J. (1991). Plant Science 74: 115-126.

14. Williams, E. G; Maheswaran, G. and Hutschinson, J. F. (1987). Plant Breeding Reviews 5: 181-236.

15. Willing, R. R. and Pryor, L. D. (1976). Theor. Appl. Genet. 47: 141-151.

Colour Plates

Chapter 2: Chromosomal Theory of Inheritance

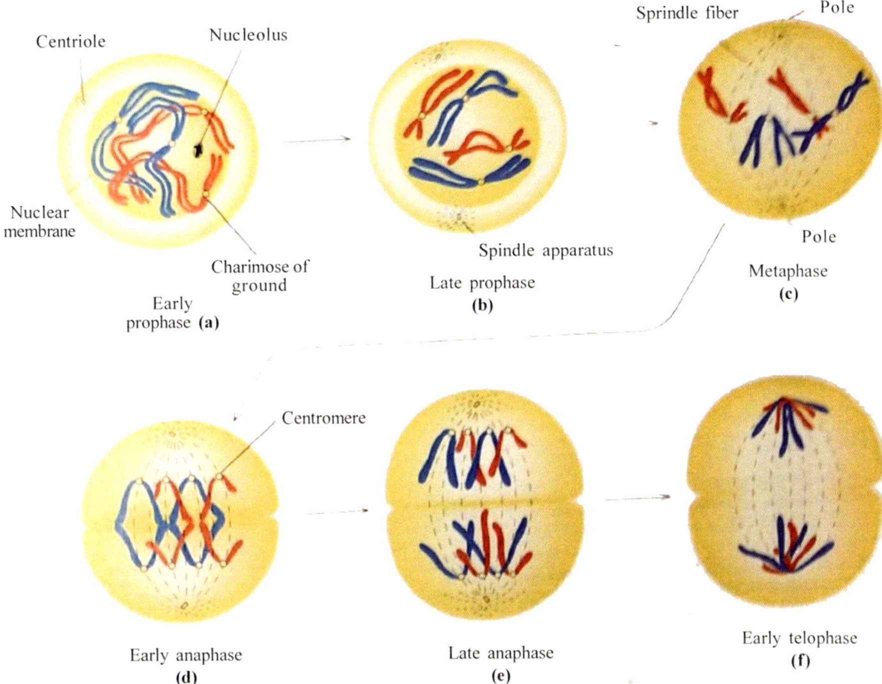

Fig. 9: Hypothetical mitotic stages with 2 pairs of chromosome

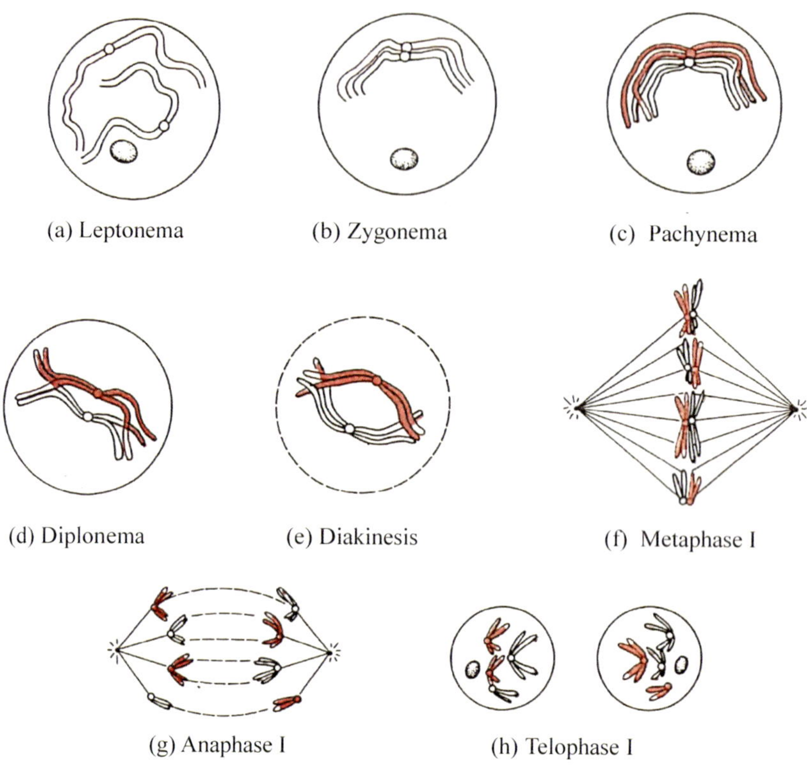

Fig. 14: Diagramatic representation of various meiotic stages of meiosis I.

Index

D

E

G

H

I&J

K

L

M

N

O

P&Q

R

S

T

U&V

W, X,Y&Z